Succeed

Eureka Math®
Grade 3
Modules 5–7

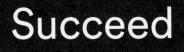

Published by Great Minds®.

Copyright © 2018 Great Minds®.

Printed in the U.S.A.

This book may be purchased from the publisher at eureka-math.org.

BAB   10   9   8   7   6   5   4   3   2

ISBN 978-1-64054-088-0

G3-M5-M7-S-06.2018

Learn ◆ Practice ◆ Succeed

Eureka Math® student materials for *A Story of Units*® (K–5) are available in the *Learn, Practice, Succeed* trio. This series supports differentiation and remediation while keeping student materials organized and accessible. Educators will find that the *Learn, Practice,* and *Succeed* series also offers coherent—and therefore, more effective—resources for Response to Intervention (RTI), extra practice, and summer learning.

Learn

Eureka Math Learn serves as a student's in-class companion where they show their thinking, share what they know, and watch their knowledge build every day. *Learn* assembles the daily classwork—Application Problems, Exit Tickets, Problem Sets, templates—in an easily stored and navigated volume.

Practice

Each *Eureka Math* lesson begins with a series of energetic, joyous fluency activities, including those found in *Eureka Math Practice.* Students who are fluent in their math facts can master more material more deeply. With *Practice,* students build competence in newly acquired skills and reinforce previous learning in preparation for the next lesson.

Together, *Learn* and *Practice* provide all the print materials students will use for their core math instruction.

Succeed

Eureka Math Succeed enables students to work individually toward mastery. These additional problem sets align lesson by lesson with classroom instruction, making them ideal for use as homework or extra practice. Each problem set is accompanied by a Homework Helper, a set of worked examples that illustrate how to solve similar problems.

Teachers and tutors can use *Succeed* books from prior grade levels as curriculum-consistent tools for filling gaps in foundational knowledge. Students will thrive and progress more quickly as familiar models facilitate connections to their current grade-level content.

Students, families, and educators:

Thank you for being part of the *Eureka Math®* community, where we celebrate the joy, wonder, and thrill of mathematics.

Nothing beats the satisfaction of success—the more competent students become, the greater their motivation and engagement. The *Eureka Math Succeed* book provides the guidance and extra practice students need to shore up foundational knowledge and build mastery with new material.

What is in the Succeed *book?*

Eureka Math Succeed books deliver supported practice sets that parallel the lessons of *A Story of Units®*. Each *Succeed* lesson begins with a set of worked examples, called *Homework Helpers*, that illustrate the modeling and reasoning the curriculum uses to build understanding. Next, students receive scaffolded practice through a series of problems carefully sequenced to begin from a place of confidence and add incremental complexity.

How should Succeed *be used?*

The collection of *Succeed* books can be used as differentiated instruction, practice, homework, or intervention. When coupled with *Affirm®*, *Eureka Math*'s digital assessment system, *Succeed* lessons enable educators to give targeted practice and to assess student progress. *Succeed*'s perfect alignment with the mathematical models and language used across *A Story of Units* ensures that students feel the connections and relevance to their daily instruction, whether they are working on foundational skills or getting extra practice on the current topic.

Where can I learn more about Eureka Math *resources?*

The Great Minds® team is committed to supporting students, families, and educators with an ever-growing library of resources, available at eureka-math.org. The website also offers inspiring stories of success in the *Eureka Math* community. Share your insights and accomplishments with fellow users by becoming a *Eureka Math* Champion.

Best wishes for a year filled with Eureka moments!

Jill Diniz

Jill Diniz
Director of Mathematics
Great Minds

Contents

Module 5: Fractions as Numbers on the Number Line

Topic F: Comparison, Order, and Size of Fractions

Module 6: Collecting and Displaying Data

Topic A: Generate and Analyze Categorical Data

Topic B: Generate and Analyze Measurement Data

Module 7: Geometry and Measurement Word Problems

Topic A: Solving Word Problems

Topic B: Attributes of Two-Dimensional Figures

Grade 3
Module 5

1. A beaker is full when the liquid reaches the fill line shown near the top. Estimate the amount of water in the beaker by shading the drawing as indicated.

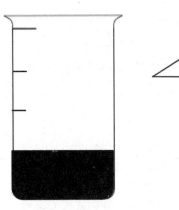

1 fourth

First, I need to partition my whole into 4 equal parts. I can estimate to draw a tick mark halfway between the top and bottom of the beaker and then make tick marks in the middle of each half. After that, I just need to shade 1 of the equal parts.

2. Juanita cut her string cheese into equal pieces as shown below. In the blank below, name the fraction of string cheese represented by the shaded part.

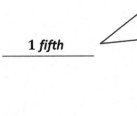

__1 fifth__

There are 5 equal parts, so each part is 1 fifth. Only 1 fifth is shaded. I can use unit form to name the fraction since I haven't learned numerical form yet.

3. In the space below, draw a small rectangle. Estimate to split it into 6 equal parts. How many lines did you draw to make 6 equal parts? What is the name of each fractional unit?

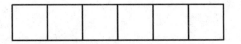

It took 5 lines to make 6 equal parts. Each fractional unit is a sixth!

To split a rectangle into 6 equal parts, I can draw a line to split it in half and then split each half into 3 equal parts. When I have 6 equal parts, my fractional unit is sixths!

Lesson 1: Specify and partition a whole into equal parts, identifying and counting unit fractions using concrete models.

© 2018 Great Minds®. eureka-math.org

3

4. Rochelle has a string that is 15 inches long. She cuts it into pieces that are each 5 inches in length. What fraction of the string is 1 piece? Use your strip from the lesson to help you. Draw a picture to show the string and how Rochelle cut it.

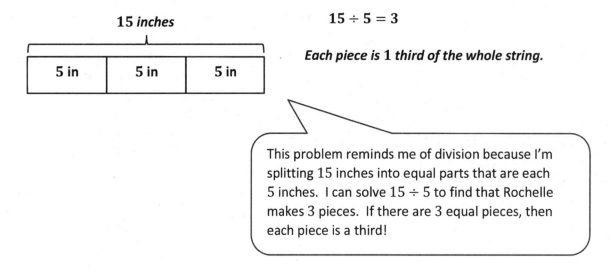

$$15 \div 5 = 3$$

Each piece is 1 third of the whole string.

This problem reminds me of division because I'm splitting 15 inches into equal parts that are each 5 inches. I can solve $15 \div 5$ to find that Rochelle makes 3 pieces. If there are 3 equal pieces, then each piece is a third!

Lesson 1:   Specify and partition a whole into equal parts, identifying and
                          counting unit fractions using concrete models.

© 2018 Great Minds®. eureka-math.org

Name _____ Date _____

1. A beaker is considered full when the liquid reaches the fill line shown near the top. Estimate the amount of water in the beaker by shading the drawing as indicated. The first one is done for you.

2. Danielle cut her candy bar into equal pieces as shown in the rectangles below. In the blanks below, name the fraction of candy bar represented by the shaded part.

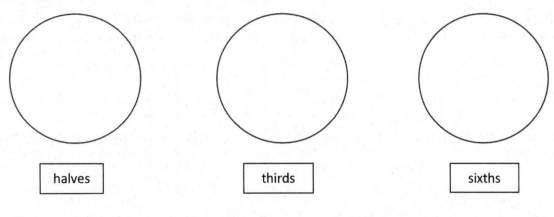

_____   _____   _____

3. Each circle represents 1 whole pie. Estimate to show how you would cut the pie into fractional units as indicated below.

EUREKA MATH

Lesson 1:   Specify and partition a whole into equal parts, identifying and counting unit fractions using concrete models.

© 2018 Great Minds®. eureka-math.org

5

4. Each rectangle represents 1 sheet of paper. Estimate to draw lines to show how you would cut the paper into fractional units as indicated below.

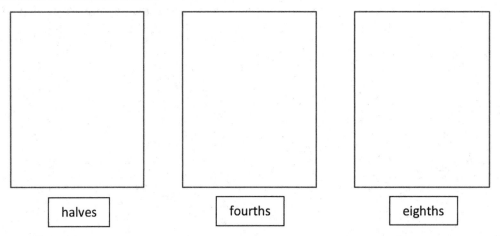

halves   fourths   eighths

5. Each rectangle represents 1 sheet of paper. Estimate to draw lines to show how you would cut the paper into fractional units as indicated below.

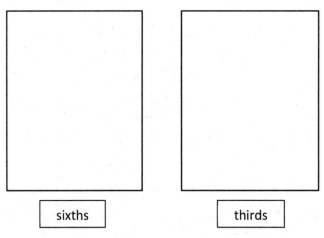

sixths   thirds

6. Yuri has a rope 12 meters long. He cuts it into pieces that are each 2 meters long. What fraction of the rope is one piece? Draw a picture. (You might fold a strip of paper to help you model the problem.)

7. Dawn bought 12 grams of chocolate. She ate half of the chocolate. How many grams of chocolate did she eat?

Lesson 1:   Specify and partition a whole into equal parts, identifying and counting unit fractions using concrete models.

© 2018 Great Minds®. eureka-math.org

1. Circle the strip that is folded to make equal parts.

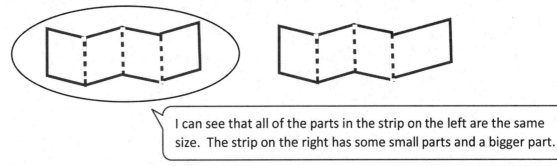

I can see that all of the parts in the strip on the left are the same size. The strip on the right has some small parts and a bigger part.

2. Dylan plans to eat 1 fourth of his candy bar. His 3 friends want him to share the rest equally. Show how Dylan and his friends can each get an equal share of the candy bar.

Dylan's friends' pieces

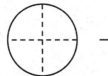

Dylan's piece

I know that 4 people are sharing the candy bar. I'll draw a fraction strip to represent the candy bar and split it into fourths. I can label Dylan's piece and the pieces that his friends will eat.

3. Nasir baked a pie and cut it into fourths. He then cut each piece in half.

   a. What fraction of the whole pie does each piece represent?

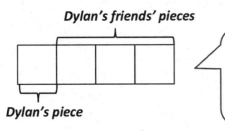

   Cut into fourths   **Each piece cut in half**

   Each piece represents 1 eighth of the whole pie.

   First, I should draw the pie and split it into 4 equal pieces. Then, I need to cut each part in half. Once I do that, I see that each piece is an eighth!

   b. Nasir ate 1 piece of pie on Tuesday and 2 pieces on Wednesday. What fraction of the whole pie was NOT eaten?

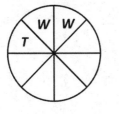

   Five eighths of the whole pie was not eaten.

   I can draw the pie and label the pieces Nasir ate. He ate 3 out of the 8 pieces, so 5 are left. So, 5 eighths of Nasir's pie is left!

EUREKA MATH

Lesson 2:   Specify and partition a whole into equal parts, identifying and counting unit fractions by folding fraction strips.

© 2018 Great Minds®. eureka-math.org

7

Name _____   Date _____

1. Circle the strips that are cut into equal parts.

2.

    a. There are _____ equal parts in all. _____ is shaded.

    b. There are _____ equal parts in all. _____ is shaded.

    c. There are _____ equal parts in all. _____ is shaded.

    d. There are _____ equal parts in all. _____ are shaded.

Lesson 2:   Specify and partition a whole into equal parts, identifying and
             counting unit fractions by folding fraction strips.

© 2018 Great Minds®. eureka-math.org

9

3. Dylan plans to eat 1 fifth of his candy bar. His 4 friends want him to share the rest equally. Show how Dylan and his friends can each get an equal share of the candy bar.

4. Nasir baked a pie and cut it in fourths. He then cut each piece in half.

    a. What fraction of the original pie does each piece represent?

    b. Nasir ate 1 piece of pie on Tuesday and 2 pieces on Wednesday. What fraction of the original pie was not eaten?

Lesson 2:   Specify and partition a whole into equal parts, identifying and
               counting unit fractions by folding fraction strips.

1. Each shape is 1 whole. Estimate to divide each into equal parts. Divide each whole using a different fractional unit. Write the name of the fractional unit on the line below the shape.

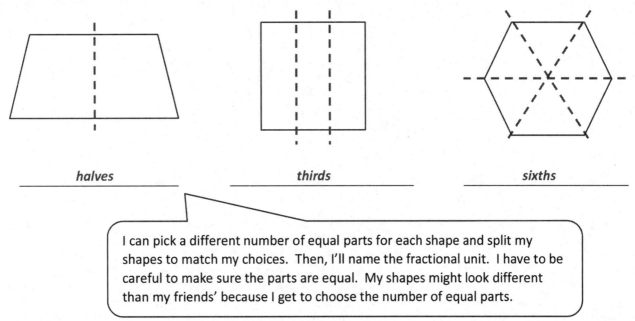

halves   _thirds_   _sixths_

> I can pick a different number of equal parts for each shape and split my shapes to match my choices. Then, I'll name the fractional unit. I have to be careful to make sure the parts are equal. My shapes might look different than my friends' because I get to choose the number of equal parts.

2. Anita uses a whole piece of paper to make a chart showing the school days in 1 week. She draws equal-sized boxes to represent each day. Draw a picture to show a possible chart. What fraction of the chart does each day take up?

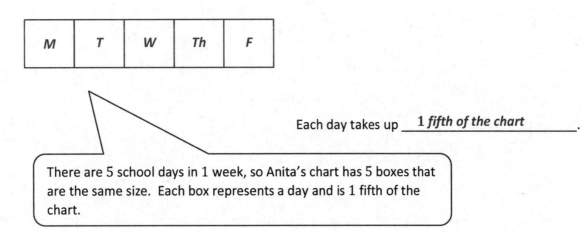

Each day takes up ____ **1 _fifth of the chart_** ____.

> There are 5 school days in 1 week, so Anita's chart has 5 boxes that are the same size. Each box represents a day and is 1 fifth of the chart.

      Lesson 3:   Specify and partition a whole into equal parts, identifying and counting unit fractions by drawing pictorial area models.   11

© 2018 Great Minds®. eureka-math.org

Name _____   Date _____

1. Each shape is a whole divided into equal parts. Name the fractional unit, and then count and tell how many of those units are shaded. The first one is done for you.

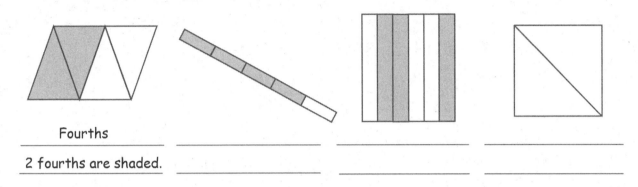

____Fourths____

2 fourths are shaded.

_____   _____   _____

2. Each shape is 1 whole. Estimate to divide each into equal parts. Divide each whole using a different fractional unit. Write the name of the fractional unit on the line below the shape.

_____   _____   _____

3. Anita uses 1 sheet of paper to make a calendar showing each month of the year. Draw Anita's calendar. Show how she can divide her calendar so that each month is given the same space. What fraction of the calendar does each month receive?

Each month receives _____.

Lesson 3:   Specify and partition a whole into equal parts, identifying and
counting unit fractions by drawing pictorial area models.

© 2018 Great Minds®. eureka-math.org

13

1. Each shape is 1 whole. Estimate to equally partition the shape, and shade to show the given fraction.

1 half

A   B

> I know that the fraction is 1 half, so I can split each shape into 2 equal parts. Then, I'll shade 1 part in each shape.

2. Each shape represents 1 whole. Match each shape to its fraction.

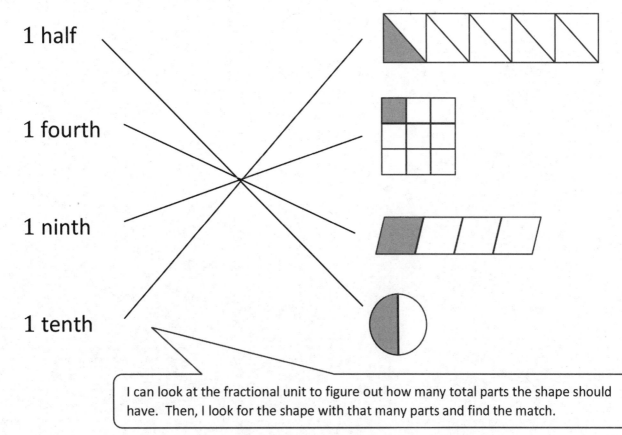

1 half

1 fourth

1 ninth

1 tenth

> I can look at the fractional unit to figure out how many total parts the shape should have. Then, I look for the shape with that many parts and find the match.

Name _____   Date _____

Each shape is 1 whole. Estimate to equally partition the shape and shade to show the given fraction.

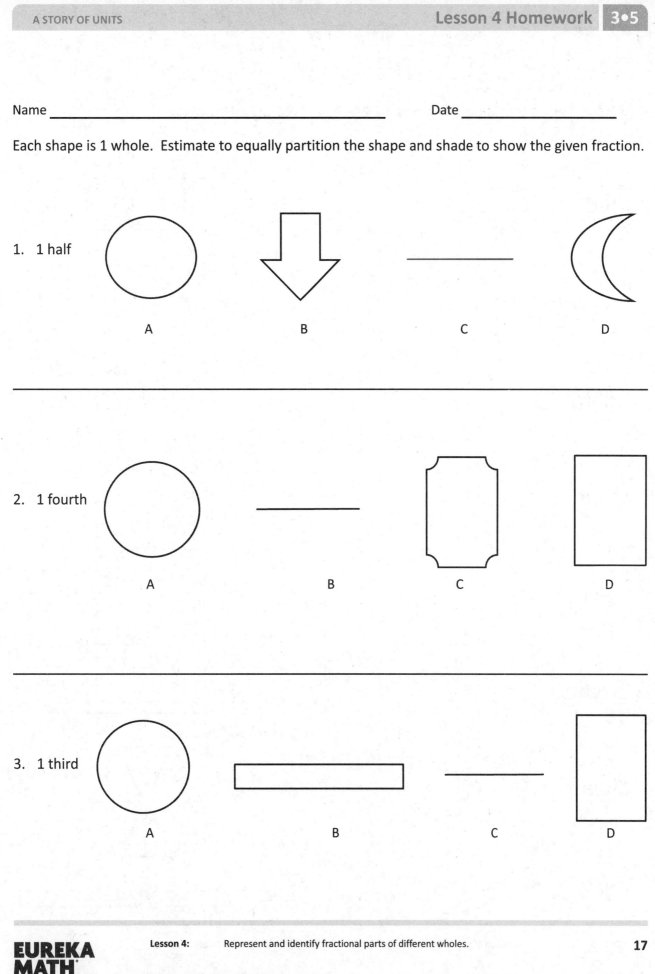

1. 1 half

        A   B   C   D

2. 1 fourth

        A   B   C   D

3. 1 third

        A   B   C   D

4. Each of the shapes represents 1 whole. Match each shape to its fraction.

1 fifth

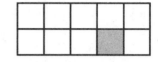

1 twelfth

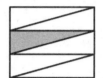

1 third

1 fourth

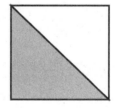

1 half

1 eighth

1 tenth

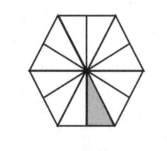

1 sixth

EUREKA
MATH®

1. Fill in the chart. Then, whisper the fractional unit.

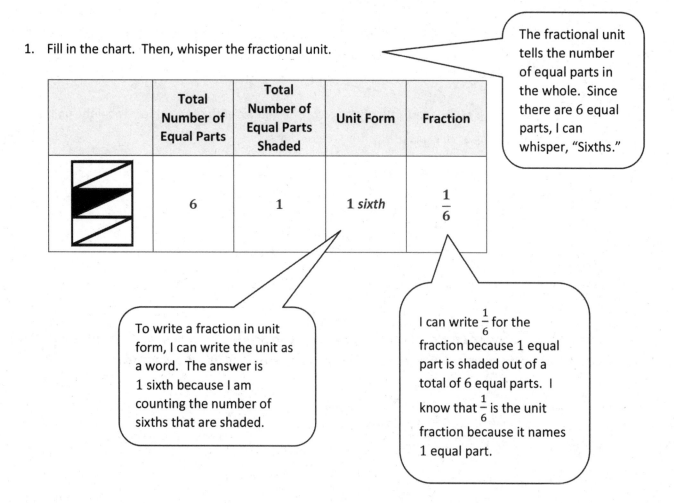

	Total Number of Equal Parts	Total Number of Equal Parts Shaded	Unit Form	Fraction
	6	1	1 *sixth*	$\dfrac{1}{6}$

> The fractional unit tells the number of equal parts in the whole. Since there are 6 equal parts, I can whisper, "Sixths."

> To write a fraction in unit form, I can write the unit as a word. The answer is 1 sixth because I am counting the number of sixths that are shaded.

> I can write $\dfrac{1}{6}$ for the fraction because 1 equal part is shaded out of a total of 6 equal parts. I know that $\dfrac{1}{6}$ is the unit fraction because it names 1 equal part.

Lesson 5:   Partition a whole into equal parts and define the equal parts to identify the unit fraction numerically.

© 2018 Great Minds®. eureka-math.org

> If 1 fifth is shaded, then that rectangle must be partitioned into 5 equal parts (fifths). The other rectangle must be partitioned into 8 equal parts (eighths).

2. Draw two identical rectangles. Shade 1 fifth of one rectangle and 1 eighth of the other. Label the unit fractions. Use your rectangles to explain why $\frac{1}{5}$ is greater than $\frac{1}{8}$.

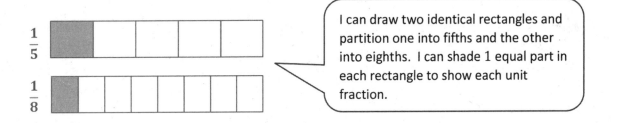

> I can draw two identical rectangles and partition one into fifths and the other into eighths. I can shade 1 equal part in each rectangle to show each unit fraction.

$\frac{1}{5}$ is greater than $\frac{1}{8}$ because both rectangles have 1 equal part shaded, but when the rectangle is cut into 5 equal parts, the parts are bigger than when the rectangle is cut into 8 equal parts.

Lesson 5:   Partition a whole into equal parts and define the equal parts to
identify the unit fraction numerically.

Name _____   Date _____

1. Fill in the chart. Each image is one whole.

	Total Number of Equal Parts	Total Number of Equal Parts Shaded	Unit Form	Fraction Form
a.				
b.				
c.				
d.				
e.				

Lesson 5:   Partition a whole into equal parts and define the equal parts to identify the unit fraction numerically.

© 2018 Great Minds®. eureka-math.org

21

2. This figure is divided into 6 parts. Are they sixths? Explain your answer.

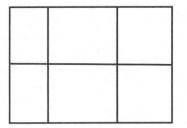

3. Terry and his 3 friends baked a pizza during his sleepover. They want to share the pizza equally. Show how Terry can slice the pizza so that he and his 3 friends can each get an equal amount with none left over.

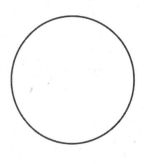

4. Draw two identical rectangles. Shade 1 seventh of one rectangle and 1 tenth of the other. Label the unit fractions. Use your rectangles to explain why $\frac{1}{7}$ is greater than $\frac{1}{10}$.

1. Complete the number sentence. Estimate to partition each strip equally, write the unit fraction inside each unit, and shade the answer.

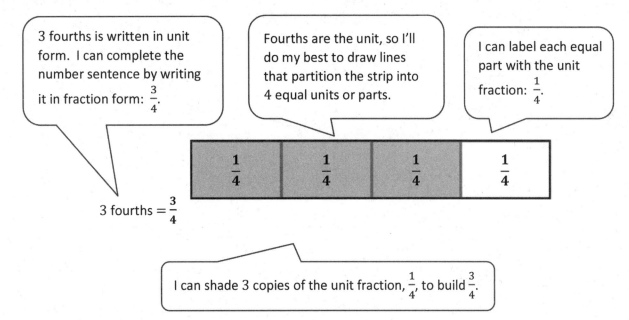

3 fourths is written in unit form. I can complete the number sentence by writing it in fraction form: $\frac{3}{4}$.

Fourths are the unit, so I'll do my best to draw lines that partition the strip into 4 equal units or parts.

I can label each equal part with the unit fraction: $\frac{1}{4}$.

3 fourths $= \frac{3}{4}$

I can shade 3 copies of the unit fraction, $\frac{1}{4}$, to build $\frac{3}{4}$.

2. Mr. Stevens buys 8 liters of soda for a party. His guests drink 1 of the 8 liters of soda.

   a. What fraction of the soda do his guests drink?

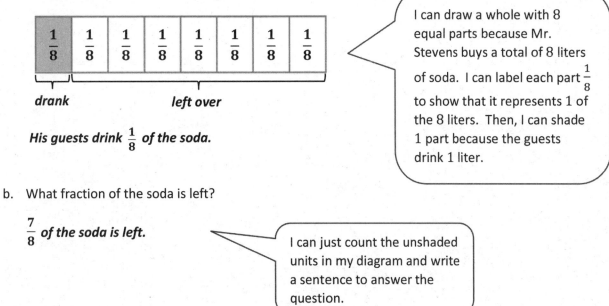

   drank   left over

   His guests drink $\frac{1}{8}$ of the soda.

I can draw a whole with 8 equal parts because Mr. Stevens buys a total of 8 liters of soda. I can label each part $\frac{1}{8}$ to show that it represents 1 of the 8 liters. Then, I can shade 1 part because the guests drink 1 liter.

   b. What fraction of the soda is left?

   $\frac{7}{8}$ of the soda is left.

I can just count the unshaded units in my diagram and write a sentence to answer the question.

Lesson 6:   Build non-unit fractions less than one whole from unit fractions.   23

Name _____   Date _____

1. Complete the number sentence. Estimate to partition each strip equally, write the unit fraction inside each unit, and shade the answer.

   Sample:

   3 fourths = $\frac{3}{4}$

$\frac{1}{4}$	$\frac{1}{4}$	$\frac{1}{4}$	$\frac{1}{4}$

   a. 2 thirds =

   b. 5 sevenths =

   c. 3 fifths =

   d. 2 eighths =

2. Mr. Abney bought 6 kilograms of rice. He cooked 1 kilogram of it for dinner.

   a. What fraction of the rice did he cook for dinner?

   b. What fraction of the rice was left?

3. Fill in the chart.

	Total Number of Equal Parts	Total Number of Shaded Equal Parts	Unit Fraction	Fraction Shaded
Sample: 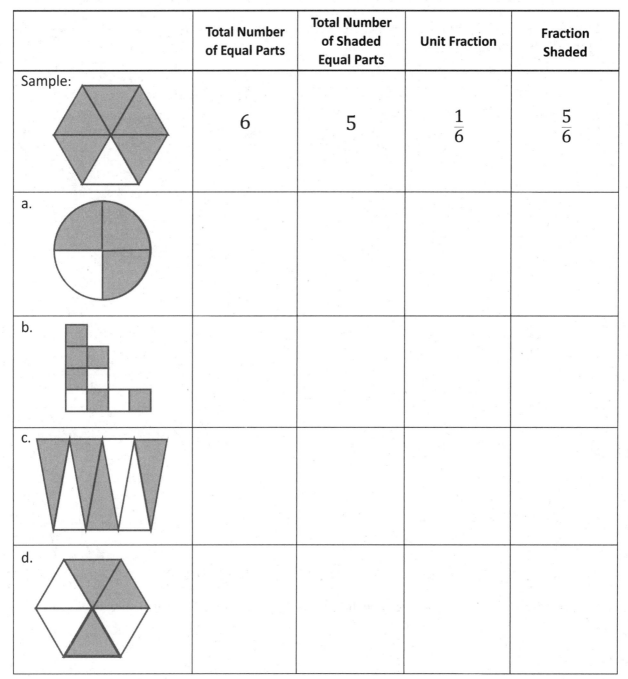	6	5	$\frac{1}{6}$	$\frac{5}{6}$
a.				
b.				
c.				
d.				

Lesson 6:   Build non-unit fractions less than one whole from unit fractions.

1. Whisper the fraction of the shape that is shaded. Then, match the shape to the amount that is not shaded.

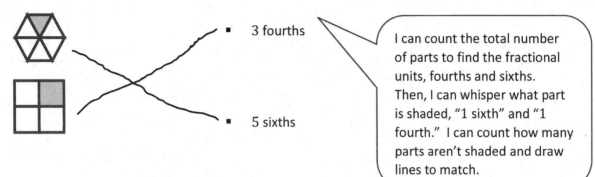

■ 3 fourths

■ 5 sixths

I can count the total number of parts to find the fractional units, fourths and sixths. Then, I can whisper what part is shaded, "1 sixth" and "1 fourth." I can count how many parts aren't shaded and draw lines to match.

2. Mom lights 10 birthday candles on the cake. Alexis blows out 9 candles. What fraction of the birthday candles are still lit? Draw and explain.

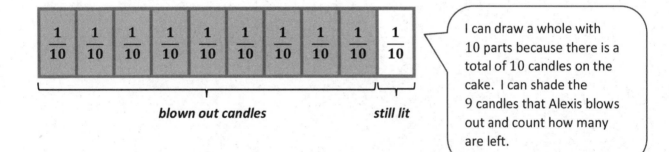

blown out candles   still lit

I can draw a whole with 10 parts because there is a total of 10 candles on the cake. I can shade the 9 candles that Alexis blows out and count how many are left.

There are a total of 10 candles, but 9 are blown out. That leaves $\frac{1}{10}$ of the candles that are still lit.

Alexis blew out all but 1 candle. Since there are 10 candles in all, the fraction of candles still lit is $\frac{1}{10}$.

Lesson 7:   Identify and represent shaded and non-shaded parts of one whole as fractions.

© 2018 Great Minds®. eureka-math.org

27

Name _____ Date _____

Whisper the fraction of the shape that is shaded. Then, match the shape to the amount that is <u>not</u> shaded.

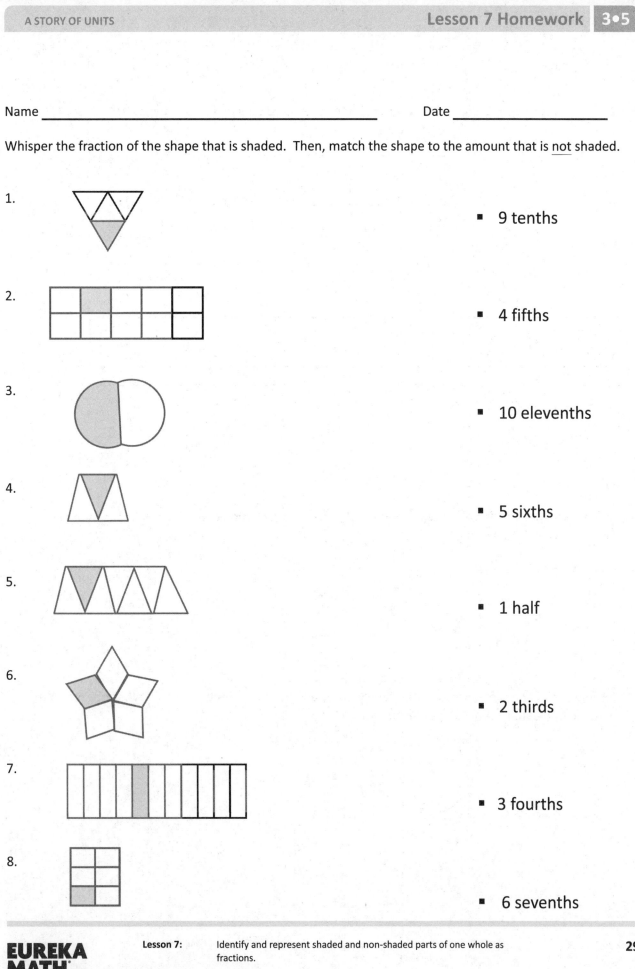

1.

2.

3.

4.

5.

6.

7.

8.

- 9 tenths

- 4 fifths

- 10 elevenths

- 5 sixths

- 1 half

- 2 thirds

- 3 fourths

- 6 sevenths

9. Each strip represents 1 whole. Write a fraction to label the shaded and unshaded parts.

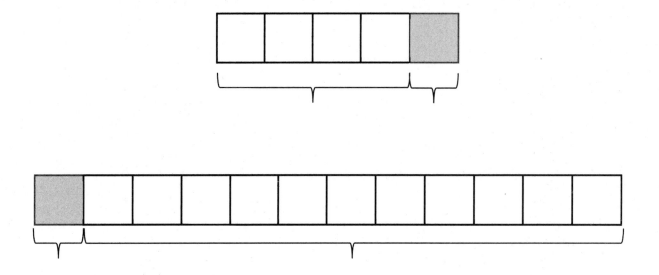

10. Carlia finished 1 fourth of her homework on Saturday. What fraction of her homework has she not finished? Draw and explain.

11. Jerome cooks 8 cups of oatmeal for his family. They eat 7 eighths of the oatmeal. What fraction of the oatmeal is uneaten? Draw and explain.

Lesson 7:   Identify and represent shaded and non-shaded parts of one whole as fractions.

1. Show a number bond representing what is shaded and unshaded in the figure. Draw a different model that would be represented by the same number bond.

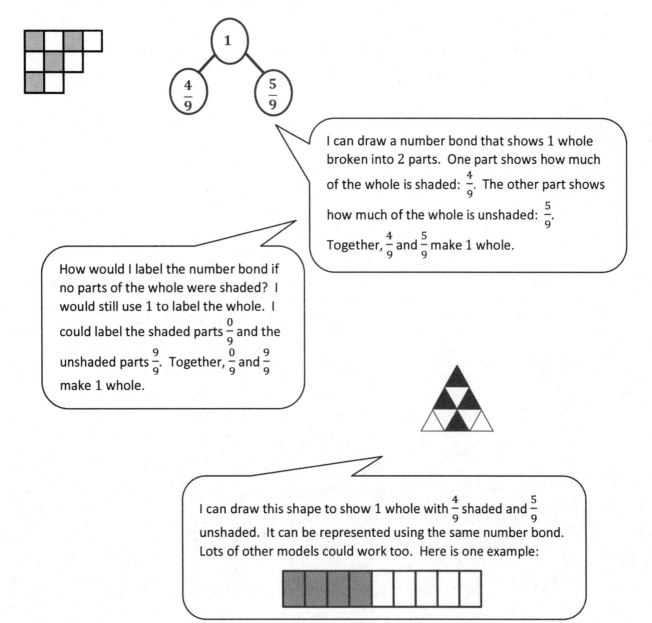

I can draw a number bond that shows 1 whole broken into 2 parts. One part shows how much of the whole is shaded: $\frac{4}{9}$. The other part shows how much of the whole is unshaded: $\frac{5}{9}$. Together, $\frac{4}{9}$ and $\frac{5}{9}$ make 1 whole.

How would I label the number bond if no parts of the whole were shaded? I would still use 1 to label the whole. I could label the shaded parts $\frac{0}{9}$ and the unshaded parts $\frac{9}{9}$. Together, $\frac{0}{9}$ and $\frac{9}{9}$ make 1 whole.

I can draw this shape to show 1 whole with $\frac{4}{9}$ shaded and $\frac{5}{9}$ unshaded. It can be represented using the same number bond. Lots of other models could work too. Here is one example:

Lesson 8:   Represent parts of one whole as fractions with number bonds.

31

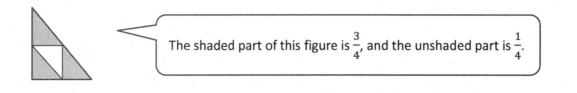

This first part is just like Problem 1.

2. Draw a number bond with 2 parts showing the shaded and unshaded fractions of each figure. Decompose both parts of the number bond into unit fractions.

The shaded part of this figure is $\frac{3}{4}$, and the unshaded part is $\frac{1}{4}$.

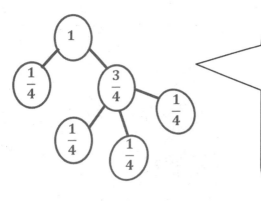

I can draw a number bond with parts of $\frac{1}{4}$ and $\frac{3}{4}$. I know that decomposing is taking apart. $\frac{1}{4}$ is already a unit fraction, but $\frac{3}{4}$ is a non-unit fraction. I can decompose $\frac{3}{4}$ into 3 copies of $\frac{1}{4}$. Now both parts of my number bond are written as unit fractions.

I can check my work by looking at all of the unit fractions. There are 4 copies of $\frac{1}{4}$, which is the same as $\frac{4}{4}$, or 1 whole.

      Lesson 8:   Represent parts of one whole as fractions with number bonds.

Name _____ Date _____

Show a number bond representing what is shaded and unshaded in each of the figures. Draw a different visual model that would be represented by the same number bond.

Sample:

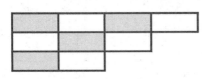 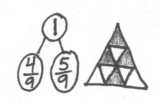

1.

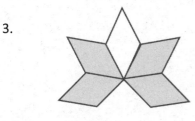

2.

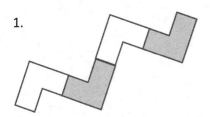

3.

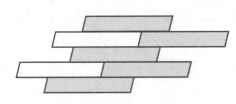

4.

Lesson 8: Represent parts of one whole as fractions with number bonds.

© 2018 Great Minds®. eureka-math.org

33

5. Draw a number bond with 2 parts showing the shaded and unshaded fractions of each figure. Decompose both parts of the number bond into unit fractions.

a. 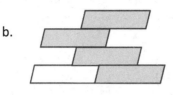   b.   c.

6. Johnny made a square peanut butter and jelly sandwich. He ate $\frac{1}{3}$ of it and left the rest on his plate. Draw a picture of Johnny's sandwich. Shade the part he left on his plate, and then draw a number bond that matches what you drew. What fraction of his sandwich did Johnny leave on his plate?

Lesson 8:   Represent parts of one whole as fractions with number bonds.

1. Each shape represents 1 whole. Fill in the chart.

> Each of these wholes is partitioned into halves. So, the unit fraction must be $\frac{1}{2}$. Three halves are shaded. I can show that by writing $\frac{3}{2}$.

	Unit Fraction	Total Number of Units Shaded	Fraction Shaded
	$\frac{1}{2}$	3	$\frac{3}{2}$

2. Estimate to draw and shade units on the fraction strips. Solve.

7 fourths $= \dfrac{7}{4}$

> 7 fourths is the unit form. I can also write it as $\frac{7}{4}$.

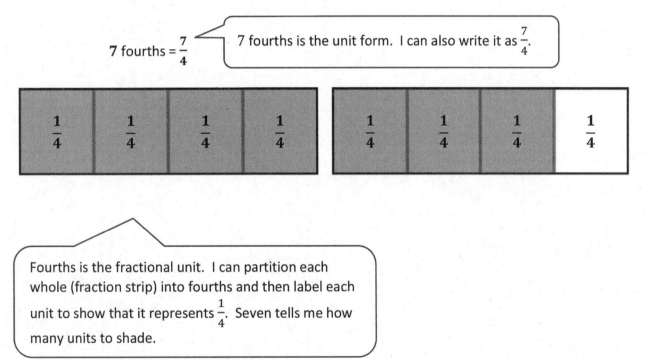

$\frac{1}{4}$	$\frac{1}{4}$	$\frac{1}{4}$	$\frac{1}{4}$

$\frac{1}{4}$	$\frac{1}{4}$	$\frac{1}{4}$	$\frac{1}{4}$

> Fourths is the fractional unit. I can partition each whole (fraction strip) into fourths and then label each unit to show that it represents $\frac{1}{4}$. Seven tells me how many units to shade.

EUREKA MATH

Lesson 9:   Build and write fractions greater than one whole using unit fractions.

35

© 2018 Great Minds®. eureka-math.org

Name _____ Date _____

1. Each shape represents 1 whole. Fill in the chart.

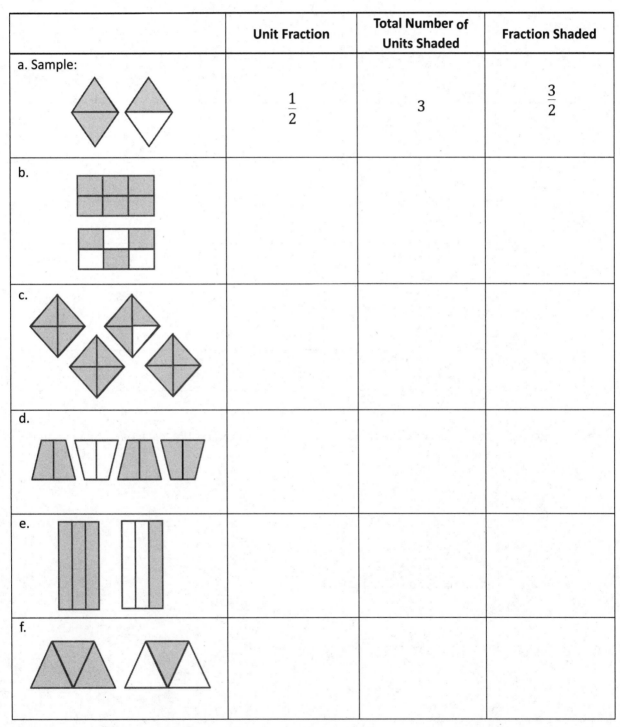

	Unit Fraction	Total Number of Units Shaded	Fraction Shaded
a. Sample:	$\frac{1}{2}$	3	$\frac{3}{2}$
b.			
c.			
d.			
e.			
f.			

2. Estimate to draw and shade units on the fraction strips. Solve.

Sample:

7 fourths = $\frac{7}{4}$

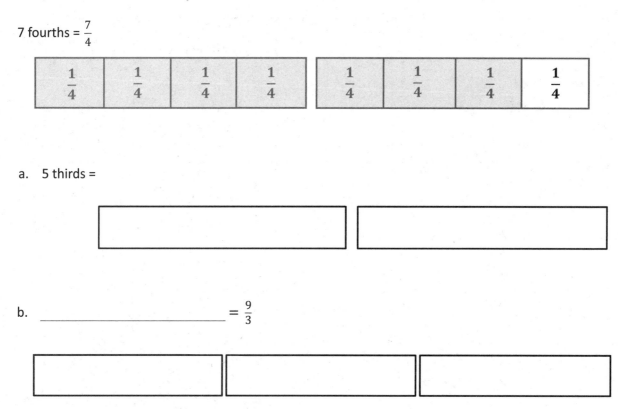

a. 5 thirds =

b. _____ = $\frac{9}{3}$

3. Reggie bought 2 candy bars. Draw the candy bars and estimate to partition each bar into 4 equal pieces.

a. Reggie ate 5 pieces. Shade the amount he ate.

b. Write a fraction to show how many candy bars Reggie ate.

1. Each fraction strip is 1 whole. The fraction strips are equal in length. Color 1 fractional unit in each strip. Then, answer the questions below.

> I can color one equal part of each whole below.

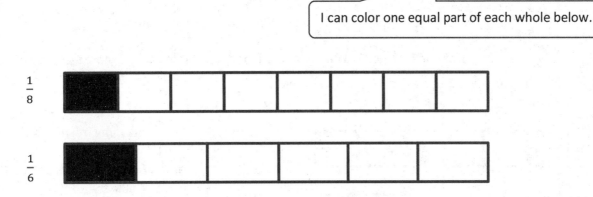

$\frac{1}{8}$

$\frac{1}{6}$

2. Circle *less than* or *greater than*. Whisper the complete sentence.

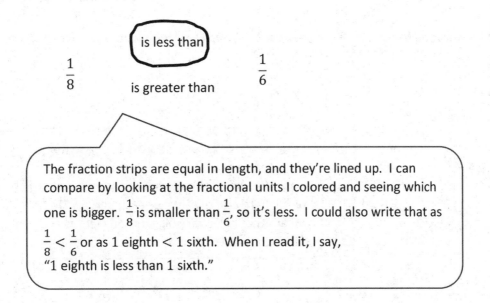

$\frac{1}{8}$ is less than $\frac{1}{6}$

is greater than

> The fraction strips are equal in length, and they're lined up. I can compare by looking at the fractional units I colored and seeing which one is bigger. $\frac{1}{8}$ is smaller than $\frac{1}{6}$, so it's less. I could also write that as $\frac{1}{8} < \frac{1}{6}$ or as 1 eighth < 1 sixth. When I read it, I say, "1 eighth is less than 1 sixth."

> I can draw fraction strips like the ones in Problem 1 to figure out which fraction is bigger.

3. Jerry feeds his dog $\frac{1}{5}$ cup of wet food and $\frac{1}{6}$ cup of dry food for dinner. Does he use more wet food or dry food? Explain your answer using pictures, numbers, and words.

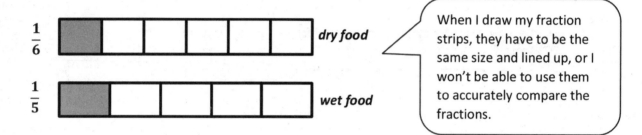

> When I draw my fraction strips, they have to be the same size and lined up, or I won't be able to use them to accurately compare the fractions.

Jerry uses more wet food because $\frac{1}{5}$ is greater than $\frac{1}{6}$. When you cut a whole into more pieces, the pieces get smaller.

4. Use >, <, or = to compare.

   a. 1 half $\left(>\right)$ $\frac{1}{8}$

   b. 1 fifth $\left(<\right)$ 1 third

> I can draw a picture to help me compare the fractions, or I can think about the size of the fractional units. I know that the more equal parts there are, the smaller each part is. That means that halves are bigger than eighths and fifths are smaller than thirds.

Lesson 10:   Compare unit fractions by reasoning about their size using fraction strips.

Name _____   Date _____

1. Each fraction strip is 1 whole. All the fraction strips are equal in length. Color 1 fractional unit in each strip. Then, answer the questions below.

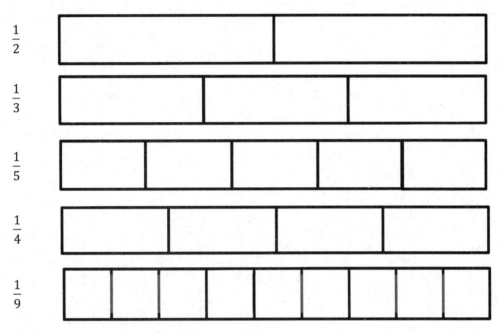

$\frac{1}{2}$

$\frac{1}{3}$

$\frac{1}{5}$

$\frac{1}{4}$

$\frac{1}{9}$

2. Circle *less than* or *greater than*. Whisper the complete sentence.

a. $\frac{1}{2}$ is less than / greater than $\frac{1}{3}$

b. $\frac{1}{9}$ is less than / greater than $\frac{1}{2}$

c. $\frac{1}{4}$ is less than / greater than $\frac{1}{2}$

d. $\frac{1}{4}$ is less than / greater than $\frac{1}{9}$

e. $\frac{1}{5}$ is less than / greater than $\frac{1}{3}$

f. $\frac{1}{5}$ is less than / greater than $\frac{1}{4}$

g. $\frac{1}{2}$ is less than / greater than $\frac{1}{5}$

h. 6 fifths is less than / greater than 3 thirds

Lesson 10:   Compare unit fractions by reasoning about their size using fraction strips.

© 2018 Great Minds®. eureka-math.org

41

3. After his football game, Malik drinks $\frac{1}{2}$ liter of water and $\frac{1}{3}$ liter of juice. Did Malik drink more water or juice? Draw and estimate to partition. Explain your answer.

4. Use >, <, or = to compare.

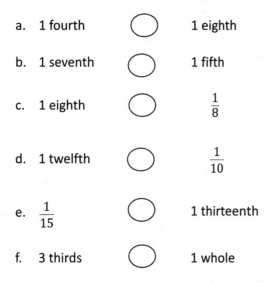

   a. 1 fourth   ◯   1 eighth

   b. 1 seventh   ◯   1 fifth

   c. 1 eighth   ◯ $\frac{1}{8}$

   d. 1 twelfth   ◯ $\frac{1}{10}$

   e. $\frac{1}{15}$ ◯   1 thirteenth

   f. 3 thirds   ◯   1 whole

5. Write a word problem about comparing fractions for your friends to solve. Be sure to show the solution so that your friends can check their work.

Lesson 10:   Compare unit fractions by reasoning about their size using fraction strips.

© 2018 Great Minds®. eureka-math.org

1. Label the unit fraction. In each blank, draw and label the same whole with a shaded unit fraction that makes the sentence true. There might be more than 1 correct way to make the sentence true.

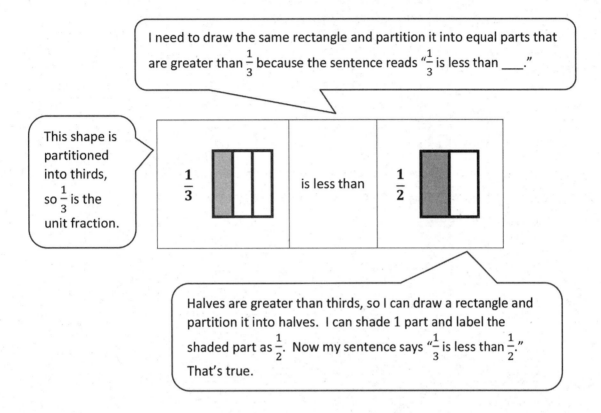

> I need to draw the same rectangle and partition it into equal parts that are greater than $\frac{1}{3}$ because the sentence reads "$\frac{1}{3}$ is less than ___."

> This shape is partitioned into thirds, so $\frac{1}{3}$ is the unit fraction.

$\frac{1}{3}$ is less than $\frac{1}{2}$

> Halves are greater than thirds, so I can draw a rectangle and partition it into halves. I can shade 1 part and label the shaded part as $\frac{1}{2}$. Now my sentence says "$\frac{1}{3}$ is less than $\frac{1}{2}$." That's true.

2. Luna drinks $\frac{1}{5}$ of a large water bottle. Gabriel drinks $\frac{1}{3}$ of a small water bottle. Gabriel says, "I drank more than you because $\frac{1}{3} > \frac{1}{5}$."

a. Use pictures and words to explain Gabriel's mistake.

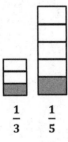

$\frac{1}{3}$ $\frac{1}{5}$

Gabriel can't compare how much water he and Luna drank. Since the wholes are different, $\frac{1}{5}$ might be bigger than $\frac{1}{3}$ like in the picture I drew.

> The important thing I notice is that the water bottles are different sizes. That means the wholes are different, so I can't compare the fractions.

EUREKA MATH

Lesson 11: Compare unit fractions with different-sized models representing the whole.

b. How could you change the problem so that Gabriel is correct? Use pictures and words to explain.

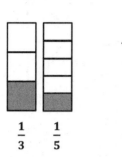

$$\frac{1}{3} \qquad \frac{1}{5}$$

I can draw models for Gabriel and Luna that are the same size. I can partition and shade the models to show $\frac{1}{3}$ and $\frac{1}{5}$. It's easy to compare the fractions now that the wholes are the same.

I could change the problem to make the wholes the same size. I could say that they both drank water from the same-sized water bottles. Then $\frac{1}{3}$ is greater than $\frac{1}{5}$. When the whole is the same, fifths are smaller than thirds.

Lesson 11: Compare unit fractions with different-sized models representing the whole.

Name _____   Date _____

Label the unit fraction. In each blank, draw and label the same whole with a shaded unit fraction that makes the sentence true. There is more than 1 correct way to make the sentence true.

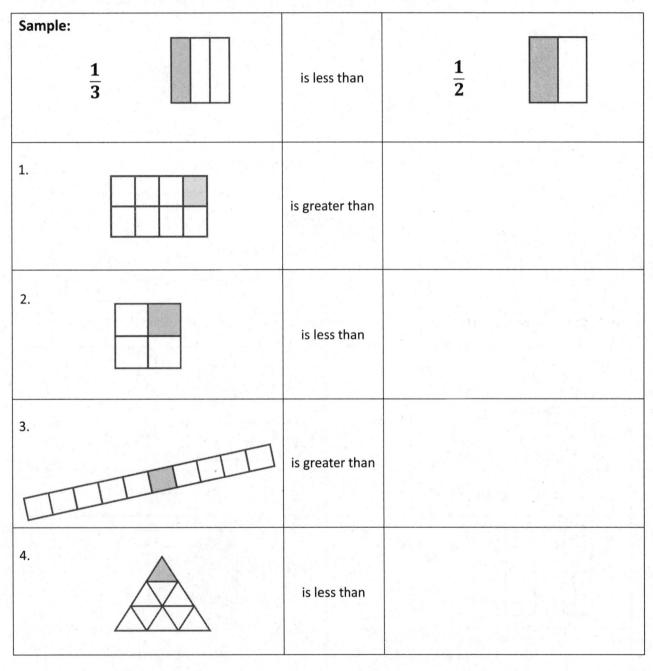

EUREKA
MATH

Lesson 11:   Compare unit fractions with different-sized models representing the whole.

© 2018 Great Minds®. eureka-math.org

45

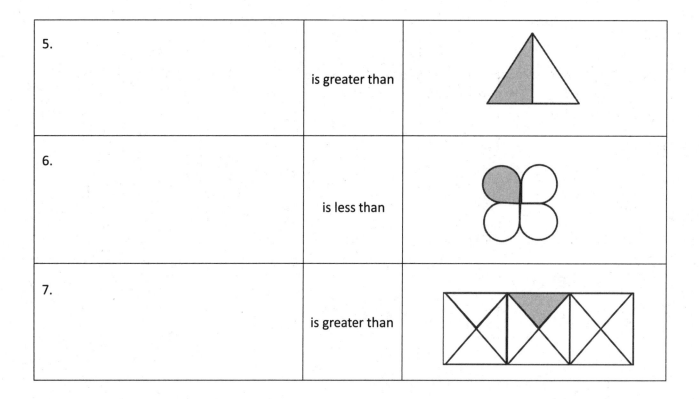

8. Fill in the blank with a fraction to make the statement true. Draw a matching model.

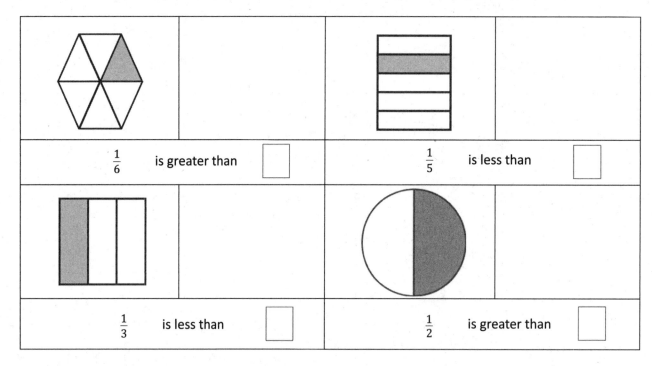

Lesson 11:   Compare unit fractions with different-sized models representing the
                whole.

© 2018 Great Minds®. eureka-math.org

9. Debbie ate $\frac{1}{8}$ of a large brownie. Julian ate $\frac{1}{2}$ of a small brownie. Julian says, "I ate more than you because $\frac{1}{2} > \frac{1}{8}$."

    a. Use pictures and words to explain Julian's mistake.

    b. How could you change the problem so that Julian is correct? Use pictures and words to explain.

1. Each shape represents the given unit fraction. Estimate to draw a possible whole. Draw a number bond that matches.

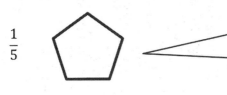

$\frac{1}{5}$

> The 5 in the fraction tells me that the unit is fifths, so there are 5 equal parts in the whole. Since this shape is a unit fraction, I can draw 5 copies of it to build my whole. There are lots of different shapes I could draw.

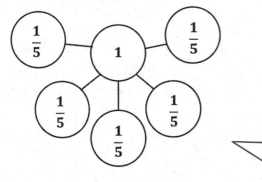

> I can make 5 copies of the unit fraction to make a whole. It's important that there are no gaps or overlaps. Overlaps would mean the parts aren't equal. If there were gaps, the whole might not be clear.

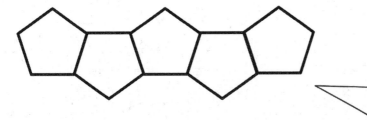

> I can draw a number bond that shows the part–whole relationship between the unit fractions and the whole. This matches the drawing because it shows that 5 copies of $\frac{1}{5}$ make a whole, or 1.

2. Cathy and Laura use this shape   to represent the unit fraction $\frac{1}{4}$. They each use

it to draw the wholes below. James says they both did it correctly. Do you agree with him? Explain
your answer.

Cathy's Shape   Laura's Shape

It looks like Cathy drew 4 copies
of the shape, but since they're
overlapping, it's really hard to
tell whether or not the parts are
equal sizes.

I can easily see in Laura's shape
that she drew 4 copies of the
shape to make a whole.

*No, I don't agree with James. Cathy's shape has a lot of overlapping, which makes it really hard to see
what the whole is. The overlapping also makes it difficult for me to see how many parts make up the
whole and whether or not the parts are equal.*

Lesson 12:   Specify the corresponding whole when presented with one equal part.

Name _____   Date _____

Each shape represents the given unit fraction. Estimate to draw a possible whole.

1. $\frac{1}{2}$

2. $\frac{1}{6}$

3. 1 third

4. 1 fourth

Lesson 12:   Specify the corresponding whole when presented with one equal part.

© 2018 Great Minds®. eureka-math.org

51

Each shape represents the given unit fraction. Estimate to draw a possible whole, label the unit fractions, and draw a number bond that matches the drawing. The first one is done for you.

5. $\frac{1}{3}$

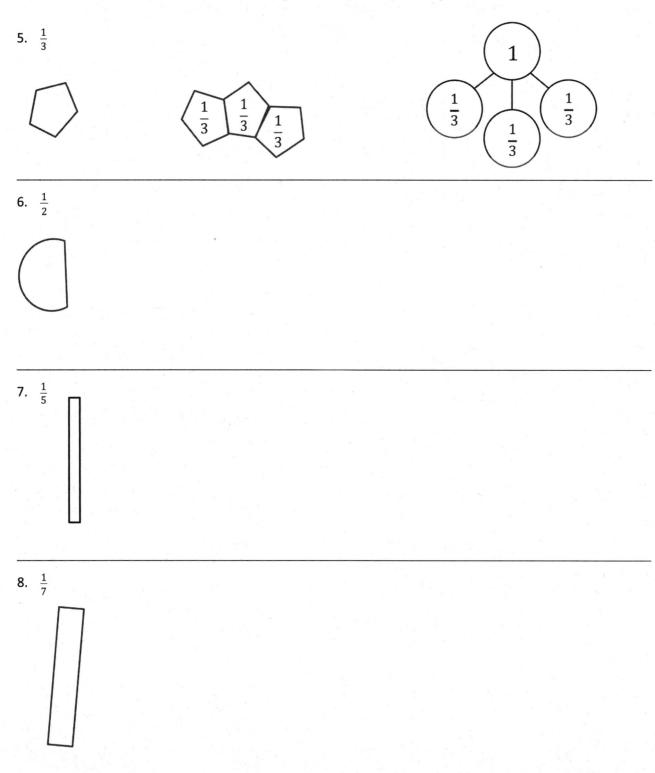

6. $\frac{1}{2}$

7. $\frac{1}{5}$

8. $\frac{1}{7}$

Lesson 12:   Specify the corresponding whole when presented with one equal part.

9. Evan and Yong used this shape , representing the unit fraction $\frac{1}{3}$, to draw 1 whole. Shania thinks both of them did it correctly. Do you agree with her? Explain your answer.

Evan's
Shape

Yong's
Shape

Lesson 12:   Specify the corresponding whole when presented with one equal part.

53

© 2018 Great Minds®. eureka-math.org

1.

The shape represents 1 whole. Write a unit fraction to describe the shaded part.	The shaded part represents 1 whole. Divide 1 whole to show the same unit fraction you wrote in part (a).
a. 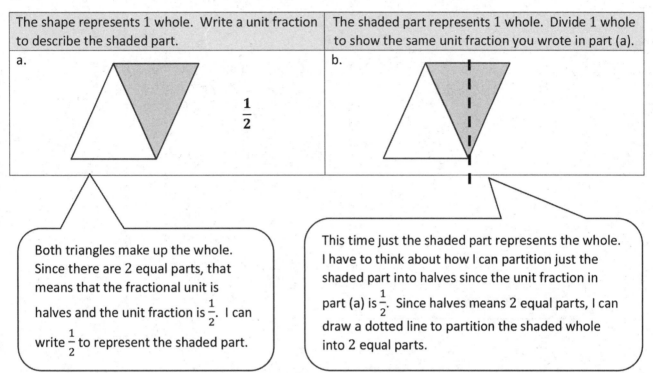 $\frac{1}{2}$	b.

Both triangles make up the whole. Since there are 2 equal parts, that means that the fractional unit is halves and the unit fraction is $\frac{1}{2}$. I can write $\frac{1}{2}$ to represent the shaded part.

This time just the shaded part represents the whole. I have to think about how I can partition just the shaded part into halves since the unit fraction in part (a) is $\frac{1}{2}$. Since halves means 2 equal parts, I can draw a dotted line to partition the shaded whole into 2 equal parts.

2.

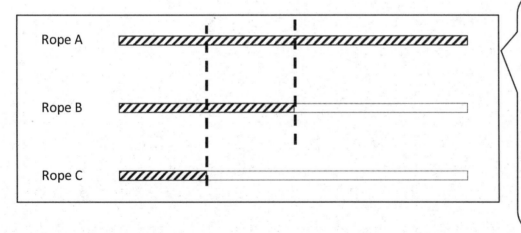

Rope A

Rope B

Rope C

I can draw a dotted line to help me compare the lengths of Ropes A and B. It looks like Rope B is about $\frac{1}{2}$ the length of Rope A. Half of 10 feet is 5 feet.

a.   If Rope A measures 10 feet long, then Rope B is about ___5___ feet long.

Lesson 13:   Identify a shaded fractional part in different ways depending on the designation of the whole.

55

© 2018 Great Minds®. eureka-math.org

b.   About how many copies of Rope C equal the length of Rope A? Draw a number bond to help you.

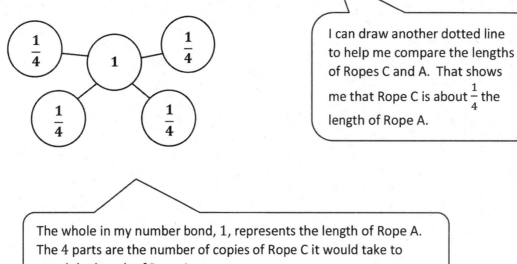

I can draw another dotted line to help me compare the lengths of Ropes C and A. That shows me that Rope C is about $\frac{1}{4}$ the length of Rope A.

The whole in my number bond, 1, represents the length of Rope A. The 4 parts are the number of copies of Rope C it would take to equal the length of Rope A.

About 4 copies of Rope C equal the length of Rope A.

      Lesson 13:   Identify a shaded fractional part in different ways depending on the designation of the whole.

Name _____   Date _____

The shape represents 1 whole. Write a fraction to describe the shaded part.	The shaded part represents 1 whole. Divide 1 whole to show the same unit fraction you wrote in Part (a).

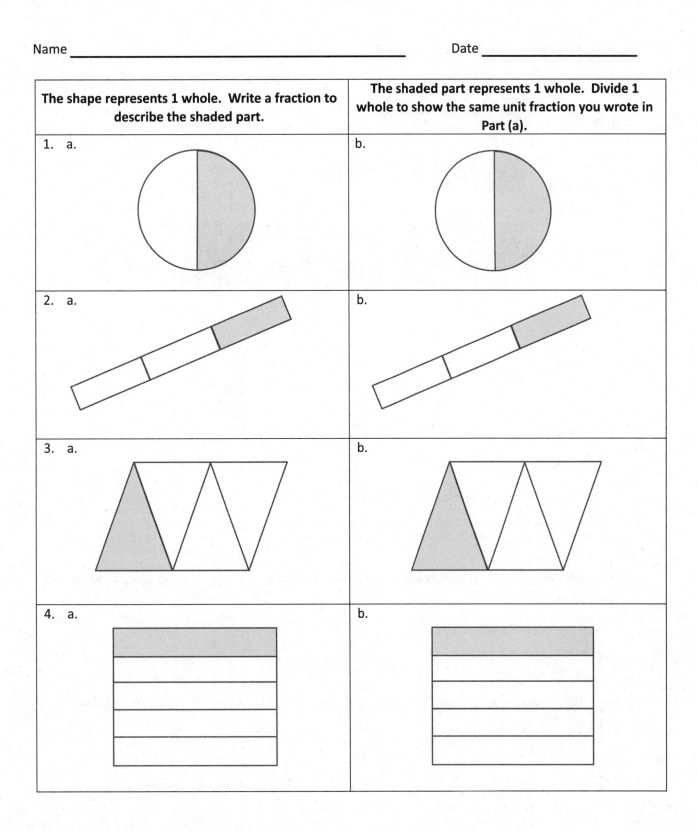

1. a.

    b.

2. a.

    b.

3. a.

    b.

4. a.

    b.

Lesson 13:   Identify a shaded fractional part in different ways depending on the designation of the whole.

© 2018 Great Minds®. eureka-math.org

57

5. Use the pictures below to complete the following statements.

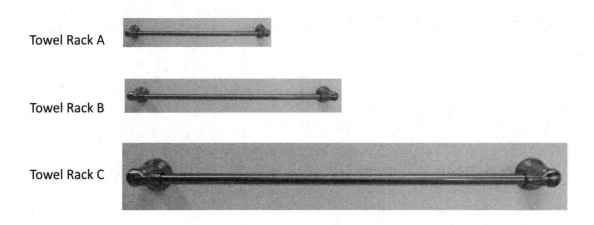

Towel Rack A

Towel Rack B

Towel Rack C

a. Towel Rack _____ is about $\frac{1}{2}$ the length of Towel Rack C.

b. Towel Rack _____ is about $\frac{1}{3}$ the length of Towel Rack C.

c. If Towel Rack C measures 6 ft long, then Towel Rack B is about _____ ft long, and Towel Rack A is about _____ ft long.

d. About how many copies of Towel Rack A equal the length of Towel Rack C? Draw number bonds to help you.

e. About how many copies of Towel Rack B equal the length of Towel Rack C? Draw number bonds to help you.

Lesson 13:　　Identify a shaded fractional part in different ways depending on the designation of the whole.

© 2018 Great Minds®. eureka-math.org

6. Draw 3 strings—B, C, and D—by following the directions below. String A is already drawn for you.

- String B is $\frac{1}{3}$ of String A.
- String C is $\frac{1}{2}$ of String B.
- String D is $\frac{1}{3}$ of String C.

Extension: String E is 5 times the length of String D.

String A

Lesson 13: Identify a shaded fractional part in different ways depending on the designation of the whole.

59

1. Draw a number bond for each fractional unit. Partition the fraction strip to show the unit fractions of the number bond. Use the fraction strip to help you label the fractions on the number line. Be sure to label the fractions at 0 and 1.

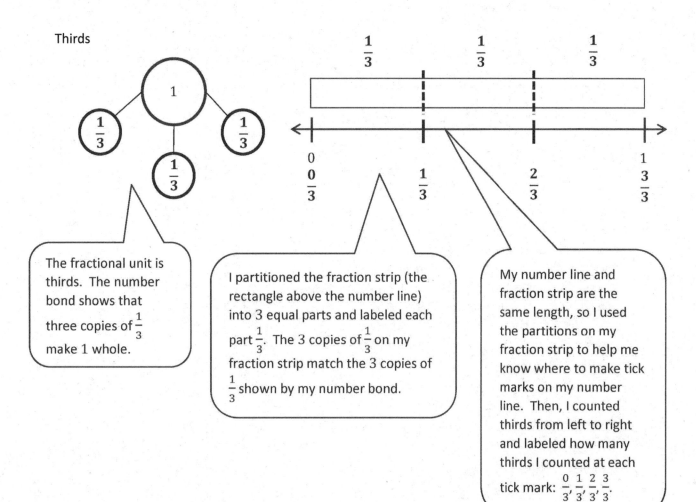

Thirds

The fractional unit is thirds. The number bond shows that three copies of $\frac{1}{3}$ make 1 whole.

I partitioned the fraction strip (the rectangle above the number line) into 3 equal parts and labeled each part $\frac{1}{3}$. The 3 copies of $\frac{1}{3}$ on my fraction strip match the 3 copies of $\frac{1}{3}$ shown by my number bond.

My number line and fraction strip are the same length, so I used the partitions on my fraction strip to help me know where to make tick marks on my number line. Then, I counted thirds from left to right and labeled how many thirds I counted at each tick mark: $\frac{0}{3}, \frac{1}{3}, \frac{2}{3}, \frac{3}{3}$.

2. A rope is 1 meter long. Mr. Lee makes a knot every $\frac{1}{4}$ meter. The first knot is at $\frac{1}{4}$ meter. The last knot is at 1 meter. Draw and label a number line from 0 meters to 1 meter to show where Mr. Lee makes knots. Label all the fractions, including 0 fourths and 4 fourths. Label 0 meters and 1 meter, too.

Mr. Lee makes knots every $\frac{1}{4}$ meter, so his rope must be partitioned into 4 equal parts.

knot   knot   knot   knot

0 m $\frac{1}{4}$ m $\frac{2}{4}$ m $\frac{3}{4}$ m   1 m

$\frac{0}{4}$ m $\frac{4}{4}$ m

rope

I can draw a number line to represent Mr. Lee's rope and then partition it into 4 equal parts. I can count by fourths from left to right starting at 0, or 0 fourths, and label them at each tick mark: 0 fourths, 1 fourth, 2 fourths, 3 fourths, 4 fourths, or 1 meter.

EUREKA
MATH®

Name _____   Date _____

1. Draw a number bond for each fractional unit. Partition the fraction strip to show the unit fractions of the number bond. Use the fraction strip to help you label the fractions on the number line. Be sure to label the fractions at 0 and 1.

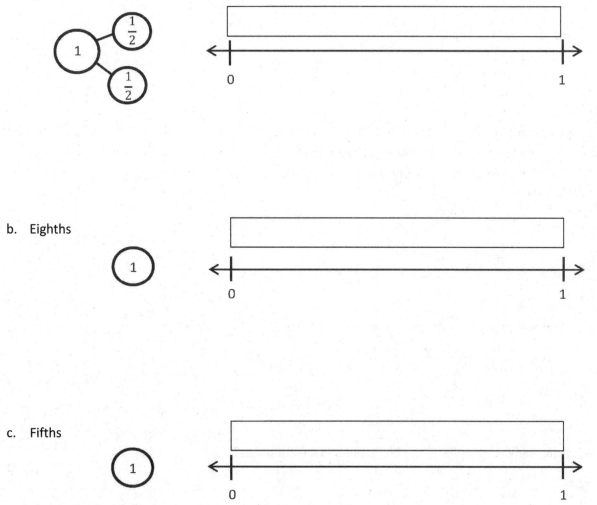

   a. Halves

   b. Eighths

   c. Fifths

Lesson 14:   Place fractions on a number line with endpoints 0 and 1.

63

© 2018 Great Minds®. eureka-math.org

2. Carter needs to wrap 7 presents. He lays the ribbon out flat and says, "If I make 6 equally spaced cuts, I'll have just enough pieces. I can use 1 piece for each package, and I won't have any pieces left over."   Does he have enough pieces to wrap all the presents?

3. Mrs. Rivera is planting flowers in her 1-meter long rectangular plant box. She divides the plant box into sections $\frac{1}{9}$ meter in length, and plants 1 seed in each section. Draw and label a fraction strip representing the plant box from 0 meters to 1 meter. Represent each section where Mrs. Rivera will plant a seed. Label all the fractions.

   a. How many seeds will she be able to plant in 1 plant box?

   b. How many seeds will she be able to plant in 4 plant boxes?

   c. Draw a number line below your fraction strip and mark all the fractions.

Lesson 14:   Place fractions on a number line with endpoints 0 and 1.

1. Estimate to label the given fraction on the number line. Be sure to label the fractions at 0 and 1. Write the fractions above the number line. Draw a number bond to match your number line.

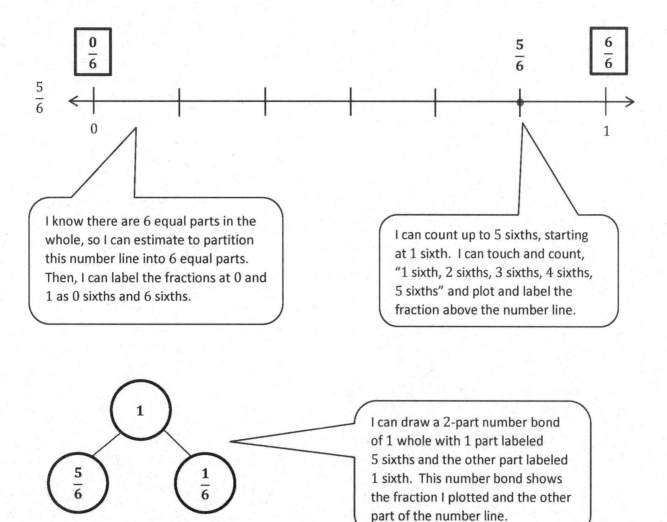

I know there are 6 equal parts in the whole, so I can estimate to partition this number line into 6 equal parts. Then, I can label the fractions at 0 and 1 as 0 sixths and 6 sixths.

I can count up to 5 sixths, starting at 1 sixth. I can touch and count, "1 sixth, 2 sixths, 3 sixths, 4 sixths, 5 sixths" and plot and label the fraction above the number line.

I can draw a 2-part number bond of 1 whole with 1 part labeled 5 sixths and the other part labeled 1 sixth. This number bond shows the fraction I plotted and the other part of the number line.

2. Claire made 6 equally spaced knots on her ribbon as shown.

$$\frac{0}{5} \qquad \frac{1}{5} \qquad \frac{2}{5} \qquad \frac{3}{5} \qquad \frac{4}{5} \qquad \frac{5}{5}$$

> I know that I need to count the number of equal parts, not the number of knots Claire made. Even though Claire made 6 knots, there are 5 equal parts.

a. Starting at the first knot and ending at the last knot, how many equal parts are formed by the 6 knots? Label each fraction at the knot.

There are 5 equal parts.

> Since there are 5 equal parts, I can label the fractions as fifths, starting with 0 fifths at the first knot and 5 fifths at the last knot.

b. What fraction of the rope is labeled at the fourth knot?

$$\frac{3}{5}$$

> I know that the first knot is 0 fifths. When I touch and count by fifths to the fourth knot, I count 3 fifths.

    Lesson 15:   Place any fraction on a number line with endpoints 0 and 1.

Name _____   Date _____

1. Estimate to label the given fractions on the number line. Be sure to label the fractions at 0 and 1. Write the fractions above the number line. Draw a number bond to match your number line. The first one is done for you.

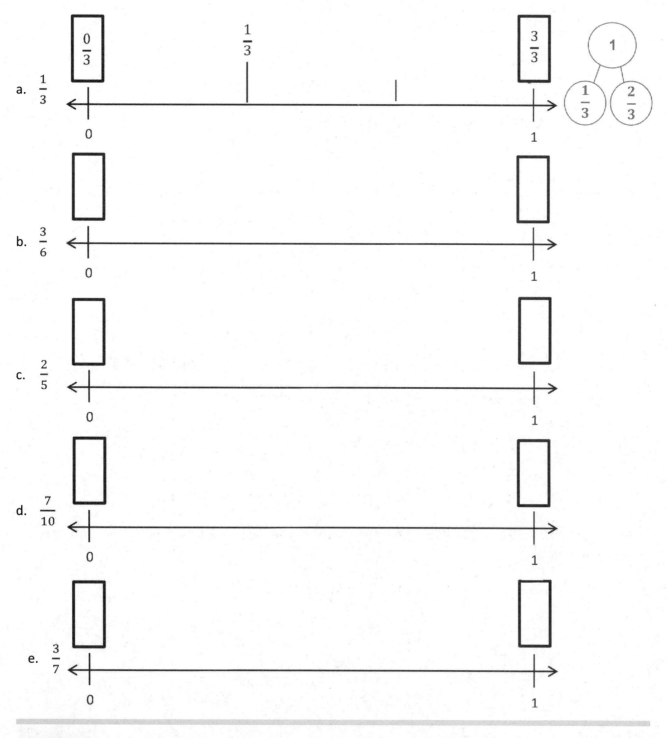

a. $\frac{1}{3}$

b. $\frac{3}{6}$

c. $\frac{2}{5}$

d. $\frac{7}{10}$

e. $\frac{3}{7}$

2. Henry has 5 dimes. Ben has 9 dimes. Tina has 2 dimes.

    a. Write the value of each person's money as a fraction of a dollar:

    Henry:

    Ben:

    Tina:

    b. Estimate to place each fraction on the number line.

$0 $1

3. Draw a number line. Use a fraction strip to locate 0 and 1. Fold the strip to make 8 equal parts.

    a. Use the strip to measure and label your number line with eighths.

    b. Count up from 0 eighths to 8 eighths on your number line. Touch each number with your finger as you count.

Lesson 15:   Place any fraction on a number line with endpoints 0 and 1.

© 2018 Great Minds®. eureka-math.org

1. Estimate to equally partition and label the fractions on the number line. Label the whole numbers as fractions, and box them.

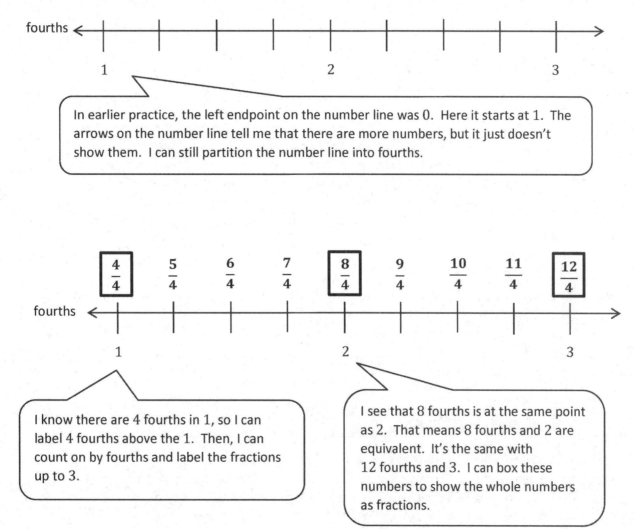

fourths

1   2   3

In earlier practice, the left endpoint on the number line was 0. Here it starts at 1. The arrows on the number line tell me that there are more numbers, but it just doesn't show them. I can still partition the number line into fourths.

$\frac{4}{4}$ $\frac{5}{4}$ $\frac{6}{4}$ $\frac{7}{4}$ $\frac{8}{4}$ $\frac{9}{4}$ $\frac{10}{4}$ $\frac{11}{4}$ $\frac{12}{4}$

fourths

1   2   3

I know there are 4 fourths in 1, so I can label 4 fourths above the 1. Then, I can count on by fourths and label the fractions up to 3.

I see that 8 fourths is at the same point as 2. That means 8 fourths and 2 are equivalent. It's the same with 12 fourths and 3. I can box these numbers to show the whole numbers as fractions.

Lesson 16:   Place whole number fractions and fractions between whole numbers on the number line.

69

2. Draw a number line with endpoints 4 and 6. Label the whole numbers. Estimate to partition each interval into sixths, and label them. Box the fractions that are located at the same points as whole numbers.

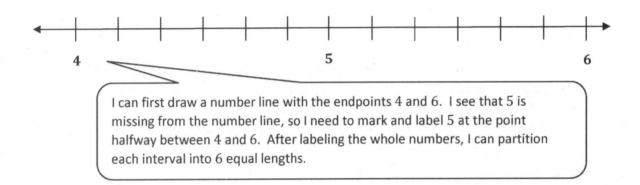

I can first draw a number line with the endpoints 4 and 6. I see that 5 is missing from the number line, so I need to mark and label 5 at the point halfway between 4 and 6. After labeling the whole numbers, I can partition each interval into 6 equal lengths.

$$\boxed{\frac{24}{6}} \quad \frac{25}{6} \quad \frac{26}{6} \quad \frac{27}{6} \quad \frac{28}{6} \quad \frac{29}{6} \quad \boxed{\frac{30}{6}} \quad \frac{31}{6} \quad \frac{32}{6} \quad \frac{33}{6} \quad \frac{34}{6} \quad \frac{35}{6} \quad \boxed{\frac{36}{6}}$$

4   5   6

This number line starts at 4. I need to figure out how many sixths are equivalent to 4. I know 6 copies of 1 sixth make 1, so 12 copies of 1 sixth make 2, 18 copies make 3, and 24 copies make 4. I notice a pattern. I am skip-counting by 6 sixths to get to the next whole number. That means I can also just multiply 4 × 6 sixths to get 24 sixths. Now that I know 24 sixths is equivalent to 4, I can count on to fill in the rest of my number line.

Lesson 16:   Place whole number fractions and fractions between whole numbers on the number line.

© 2018 Great Minds®. eureka-math.org

Name _____ Date _____

1. Estimate to equally partition and label the fractions on the number line. Label the wholes as fractions, and box them. The first one is done for you.

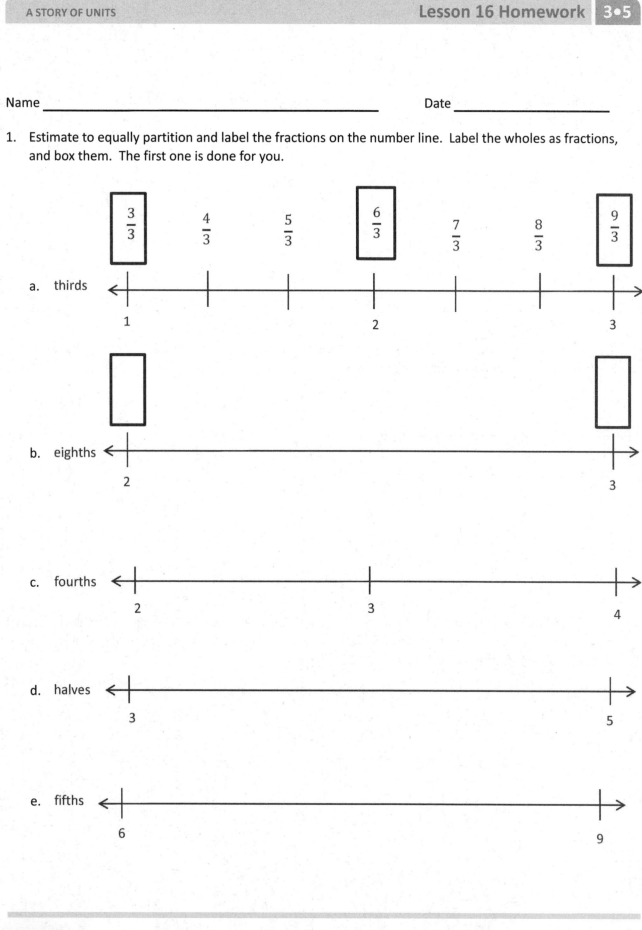

a. thirds

b. eighths

c. fourths

d. halves

e. fifths

EUREKA
MATH

Lesson 16: Place whole number fractions and fractions between whole numbers
on the number line.

© 2018 Great Minds®. eureka-math.org

71

2. Partition each whole into sixths. Label each fraction. Count up as you go. Box the fractions that are located at the same points as whole numbers.

3   4   5

3. Partition each whole into halves. Label each fraction. Count up as you go. Box the fractions that are located at the same points as whole numbers.

4   5   6   7

4. Draw a number line with endpoints 0 and 3. Label the wholes. Partition each whole into fifths. Label all the fractions from 0 to 3. Box the fractions that are located at the same points as whole numbers. Use a separate paper if you need more space.

Lesson 16:   Place whole number fractions and fractions between whole numbers on the number line.

© 2018 Great Minds®. eureka-math.org

EUREKA MATH

1. Locate and label the following fractions on the number line.

I notice that all of these fractions are thirds. That means I need to partition my number line into thirds.

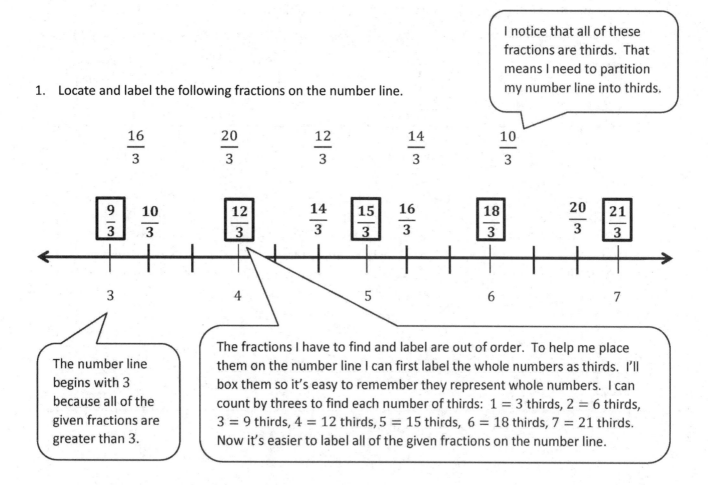

The number line begins with 3 because all of the given fractions are greater than 3.

The fractions I have to find and label are out of order. To help me place them on the number line I can first label the whole numbers as thirds. I'll box them so it's easy to remember they represent whole numbers. I can count by threes to find each number of thirds: $1 = 3$ thirds, $2 = 6$ thirds, $3 = 9$ thirds, $4 = 12$ thirds, $5 = 15$ thirds, $6 = 18$ thirds, $7 = 21$ thirds. Now it's easier to label all of the given fractions on the number line.

EUREKA MATH

2. Students measure the lengths of their earthworms in science class. Nathan's measures 3 inches long. Elisha's is $\frac{15}{4}$ inches long. Whose earthworm is longer? Draw a number line to help prove your answer.

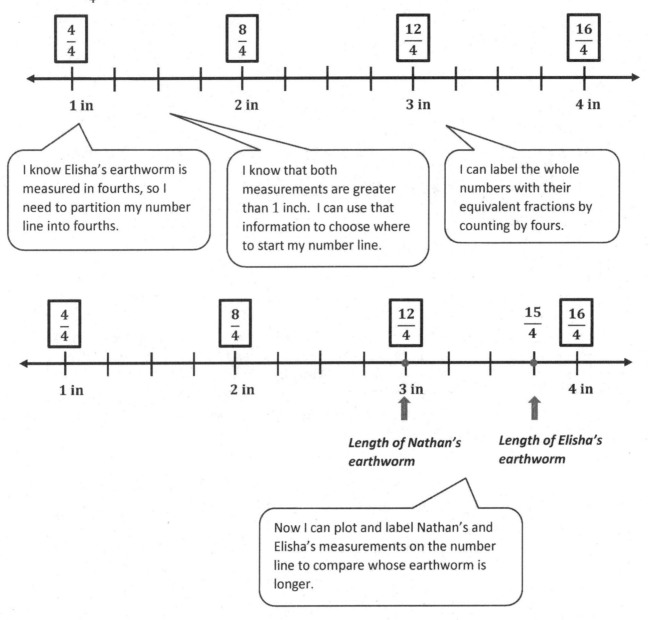

I know Elisha's earthworm is measured in fourths, so I need to partition my number line into fourths.

I know that both measurements are greater than 1 inch. I can use that information to choose where to start my number line.

I can label the whole numbers with their equivalent fractions by counting by fours.

Length of Nathan's earthworm

Length of Elisha's earthworm

Now I can plot and label Nathan's and Elisha's measurements on the number line to compare whose earthworm is longer.

Elisha's earthworm is longer. I can see that 3 inches, or $\frac{12}{4}$, comes before $\frac{15}{4}$ inches on the number line.

Lesson 17:   Practice placing various fractions on the number line.

Name _____   Date _____

1. Locate and label the following fractions on the number line.

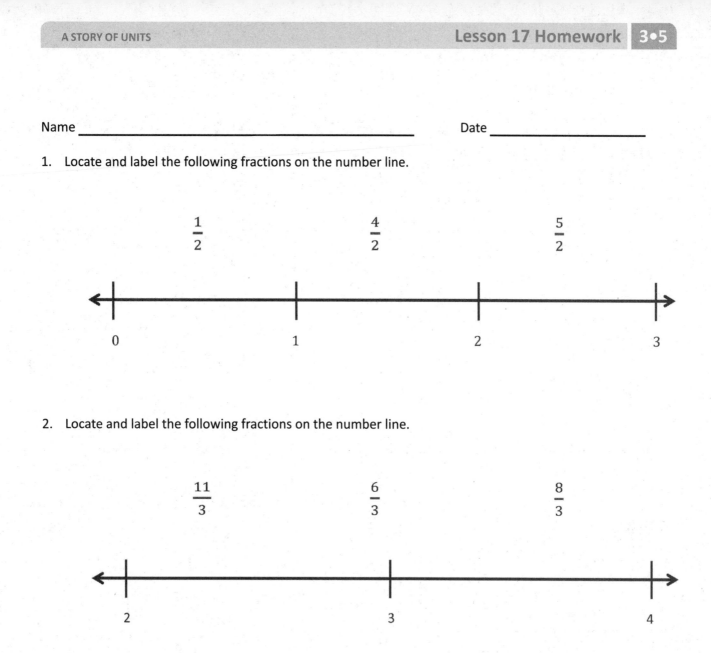

2. Locate and label the following fractions on the number line.

3. Locate and label the following fractions on the number line.

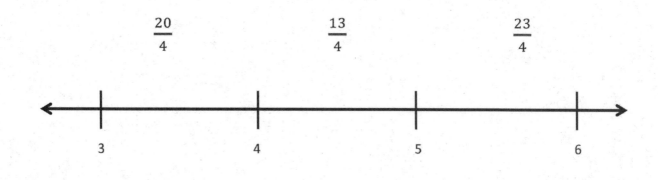

4. Wayne went on a 4-kilometer hike. He took a break at $\frac{4}{3}$ kilometers. He took a drink of water at $\frac{10}{3}$ kilometers. Show Wayne's hike on the number line. Include his starting and finishing place and the 2 points where he stopped.

5. Ali wants to buy a piano. The piano measures $\frac{19}{4}$ feet long. She has a space 5 feet long for the piano in her house. Does she have enough room? Draw a number line to show, and explain your answer.

4 ft   5 ft

Lesson 17:   Practice placing various fractions on the number line.

© 2018 Great Minds®. eureka-math.org

Place the two fractions on the number line. Circle the fraction with the distance closest to 0. Then, compare using >, <, or =.

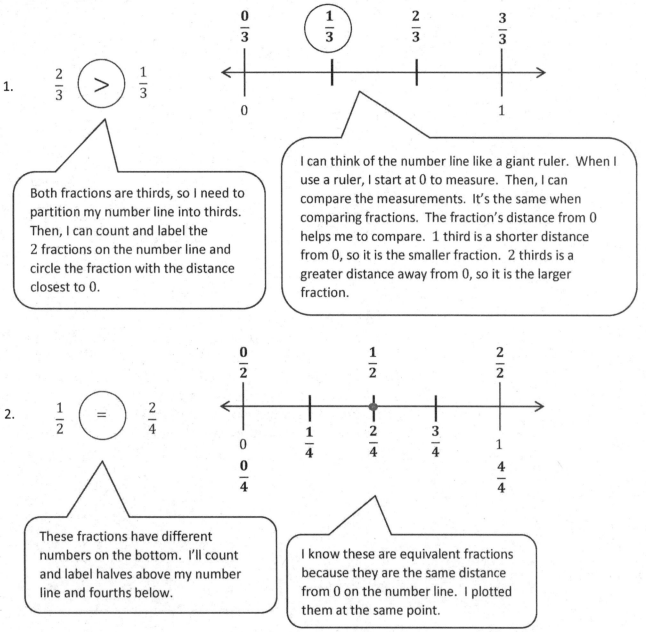

1. $\frac{2}{3}$ > $\frac{1}{3}$

Both fractions are thirds, so I need to partition my number line into thirds. Then, I can count and label the 2 fractions on the number line and circle the fraction with the distance closest to 0.

I can think of the number line like a giant ruler. When I use a ruler, I start at 0 to measure. Then, I can compare the measurements. It's the same when comparing fractions. The fraction's distance from 0 helps me to compare. 1 third is a shorter distance from 0, so it is the smaller fraction. 2 thirds is a greater distance away from 0, so it is the larger fraction.

2. $\frac{1}{2}$ = $\frac{2}{4}$

These fractions have different numbers on the bottom. I'll count and label halves above my number line and fourths below.

I know these are equivalent fractions because they are the same distance from 0 on the number line. I plotted them at the same point.

Lesson 18: Compare fractions and whole numbers on the number line by reasoning about their distance from 0.

77

© 2018 Great Minds®. eureka-math.org

3. To get to the library, John walks $\frac{1}{3}$ mile from his house. Susan walks $\frac{3}{4}$ mile from her house. Draw a number line to model how far each student walks. Who walks farther? Explain how you know using pictures, numbers, and words.

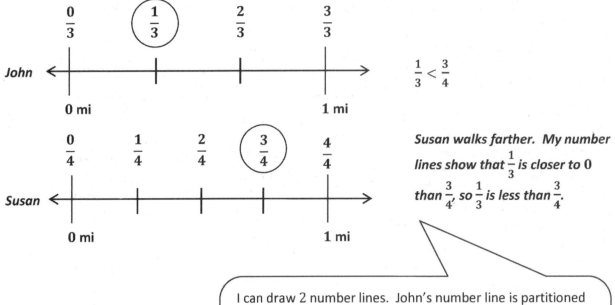

$\frac{1}{3} < \frac{3}{4}$

Susan walks farther. My number lines show that $\frac{1}{3}$ is closer to 0 than $\frac{3}{4}$, so $\frac{1}{3}$ is less than $\frac{3}{4}$.

I can draw 2 number lines. John's number line is partitioned into thirds, and Susan's number line is partitioned into fourths. I have to make sure that both my number lines have the same distance from 0 to 1 because if the whole changes, then the distance between the fractions also changes. I wouldn't be able to compare the 2 distances accurately.

Lesson 18:   Compare fractions and whole numbers on the number line by reasoning about their distance from 0.

Name _____   Date _____

Place the two fractions on the number line. Circle the fraction with the distance closest to 0. Then, compare using >, <, or =.

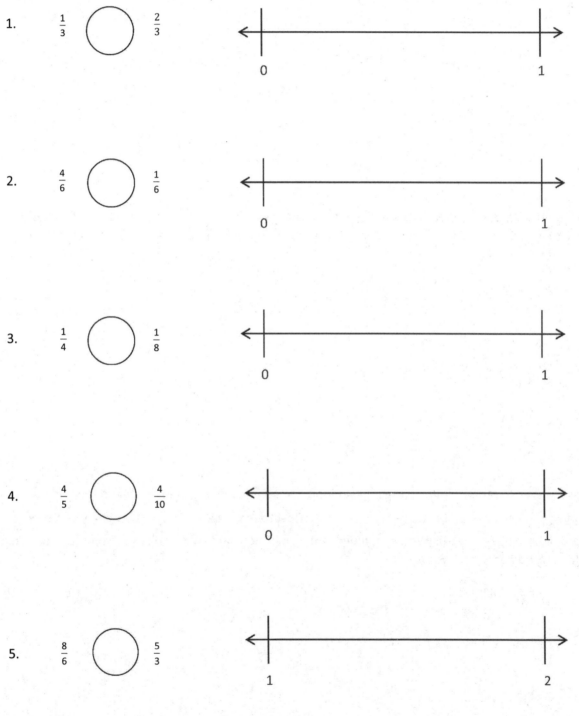

1. $\frac{1}{3}$ ◯ $\frac{2}{3}$

2. $\frac{4}{6}$ ◯ $\frac{1}{6}$

3. $\frac{1}{4}$ ◯ $\frac{1}{8}$

4. $\frac{4}{5}$ ◯ $\frac{4}{10}$

5. $\frac{8}{6}$ ◯ $\frac{5}{3}$

EUREKA
MATH®

Lesson 18:   Compare fractions and whole numbers on the number line by
reasoning about their distance from 0.

79

© 2018 Great Minds®. eureka-math.org

6. Liz and Jay each have a piece of string. Liz's string is $\frac{4}{6}$ yards long, and Jay's string is $\frac{5}{7}$ yards long.   Whose string is longer? Draw a number line to model the length of both strings. Explain the comparison using pictures, numbers, and words.

7. In a long jump competition, Wendy jumped $\frac{9}{10}$ meters, and Judy jumped $\frac{10}{9}$ meters. Draw a number line to model the distance of each girl's long jump. Who jumped the shorter distance? Explain how you know using pictures, numbers, and words.

8. Nikki has 3 pieces of yarn. The first piece is $\frac{5}{6}$ feet long, the second piece is $\frac{5}{3}$ feet long, and the third piece is $\frac{3}{2}$ feet long. She wants to arrange them from the shortest to the longest. Draw a number line to model the length of each piece of yarn. Write a number sentence using <, >, or = to compare the pieces. Explain using pictures, numbers, and words.

Lesson 18:   Compare fractions and whole numbers on the number line by reasoning about their distance from 0.

© 2018 Great Minds®. eureka-math.org

1. Divide the number line into the given fractional unit. Then, label the fractions. Write each whole number as a fraction using the given unit.

Fifths

$\frac{3}{5}$ $\frac{14}{5}$ $\frac{8}{5}$

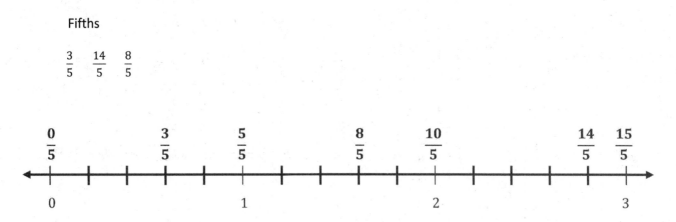

2. Use the number line above to compare the following using >, <, or =.

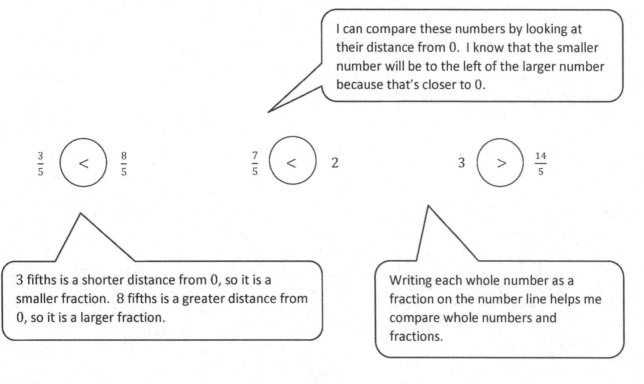

I can compare these numbers by looking at their distance from 0. I know that the smaller number will be to the left of the larger number because that's closer to 0.

$\frac{3}{5}$ (<) $\frac{8}{5}$

$\frac{7}{5}$ (<) 2

3 (>) $\frac{14}{5}$

3 fifths is a shorter distance from 0, so it is a smaller fraction. 8 fifths is a greater distance from 0, so it is a larger fraction.

Writing each whole number as a fraction on the number line helps me compare whole numbers and fractions.

Lesson 19:   Understand distance and position on the number line as strategies for comparing fractions. (Optional)

© 2018 Great Minds®. eureka-math.org

81

3. Use the number line from Problem 1 to help you. Which is larger: 2 or $\frac{9}{5}$? Use words, pictures, and numbers to explain your answer.

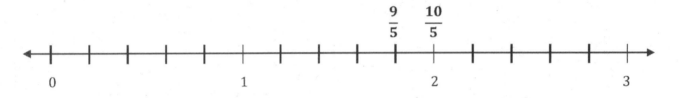

2 is larger than $\frac{9}{5}$. We can see that $\frac{9}{5}$ is to the left of 2 on the number line, which means that $\frac{9}{5}$ is closer to 0, so $\frac{9}{5}$ is less than 2.

     Lesson 19:   Understand distance and position on the number line as strategies for comparing fractions. (Optional)

Name _____   Date _____

1. Divide each number line into the given fractional unit. Then, place the fractions. Write each whole as a fraction.

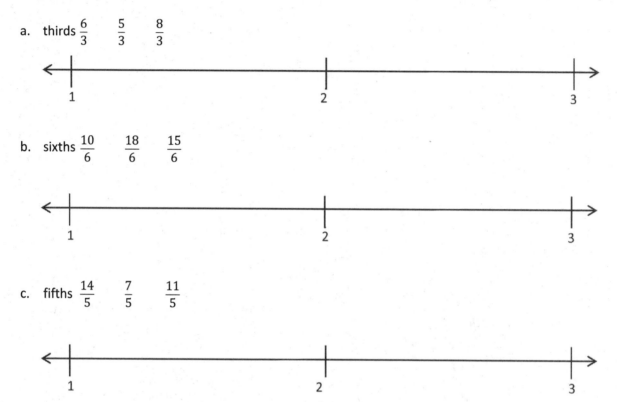

a. thirds $\frac{6}{3}$ $\frac{5}{3}$ $\frac{8}{3}$

b. sixths $\frac{10}{6}$ $\frac{18}{6}$ $\frac{15}{6}$

c. fifths $\frac{14}{5}$ $\frac{7}{5}$ $\frac{11}{5}$

2. Use the number lines above to compare the following fractions using >, <, or =.

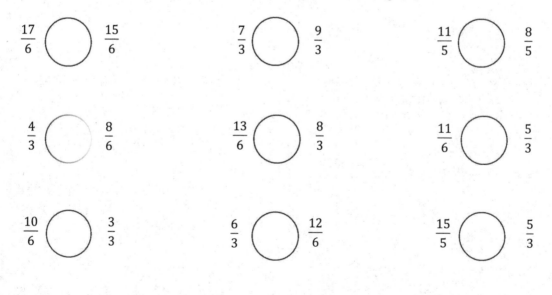

$\frac{17}{6}$ ◯ $\frac{15}{6}$ $\frac{7}{3}$ ◯ $\frac{9}{3}$ $\frac{11}{5}$ ◯ $\frac{8}{5}$

$\frac{4}{3}$ ◯ $\frac{8}{6}$ $\frac{13}{6}$ ◯ $\frac{8}{3}$ $\frac{11}{6}$ ◯ $\frac{5}{3}$

$\frac{10}{6}$ ◯ $\frac{3}{3}$ $\frac{6}{3}$ ◯ $\frac{12}{6}$ $\frac{15}{5}$ ◯ $\frac{5}{3}$

Lesson 19:   Understand distance and position on the number line as strategies for comparing fractions. (Optional)

83

© 2018 Great Minds®. eureka-math.org

3. Use fractions from the number lines in Problem 1. Complete the sentence. Use words, pictures, or numbers to explain how you made that comparison.

_____ is *greater than* _____.

4. Use fractions from the number lines in Problem 1. Complete the sentence. Use words, pictures, or numbers to explain how you made that comparison.

_____ is *less than* _____.

5. Use fractions from the number lines in Problem 1. Complete the sentence. Use words, pictures, or numbers to explain how you made that comparison.

_____ is *equal to* _____.

Lesson 19: Understand distance and position on the number line as strategies for comparing fractions. (Optional)

© 2018 Great Minds®. eureka-math.org

1. These two shapes both show $\frac{3}{4}$ shaded.

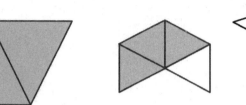

I can see that both shapes are made up of triangles, but the size of the triangles is different in each shape.

a. Are the shaded areas equivalent? Why or why not?

No, the shaded areas are not equivalent. Both shapes have 3 shaded triangles, but the size of the triangles in each shape is different. That means that the shaded areas can't be equivalent.

b. Draw two different representations of $\frac{3}{4}$ that are equivalent.

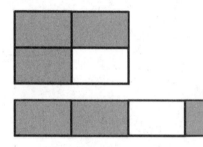

I can use the same units to draw two different representations of $\frac{3}{4}$ that are equivalent. I can rearrange the units to make a different shape.

2. Brian walked $\frac{2}{4}$ mile down the street. Wilson walked $\frac{2}{4}$ mile around the block. Who walked more? Explain your thinking.

Brian   _____

Wilson   ☐

I can see that these shapes are different, but I need to think about the units. They both walked $\frac{2}{4}$ mile, and since the units (miles) and the fractions are the same, the fractions are equivalent.

They both walked the same amount because the units are the same. They both walked $\frac{2}{4}$ mile even though they walked in different ways. Brian walked in a straight line, and Wilson walked in a rectangular shape. The shapes look different, but they are both the same distance, $\frac{2}{4}$ mile.

Lesson 20:   Recognize and show that equivalent fractions have the same size, though not necessarily the same shape.   85

© 2018 Great Minds®. eureka-math.org

Name _____   Date _____

1. Label the shaded fraction. Draw 2 different representations of the same fractional amount.

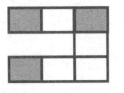

2. These two shapes both show $\frac{4}{5}$.

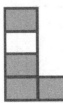

   a. Are the shapes equivalent? Why or why not?

   b. Draw two different representations of $\frac{4}{5}$ that are equivalent.

3. Diana ran a quarter mile straight down the street. Becky ran a quarter mile on a track. Who ran more? Explain your thinking.

   Diana _____

   Becky   ⬭

Lesson 20:   Recognize and show that equivalent fractions have the same size, though not necessarily the same shape.

© 2018 Great Minds®. eureka-math.org

87

1. Use the fractional units on the left to count up on the number line. Label the missing fractions on the blanks.

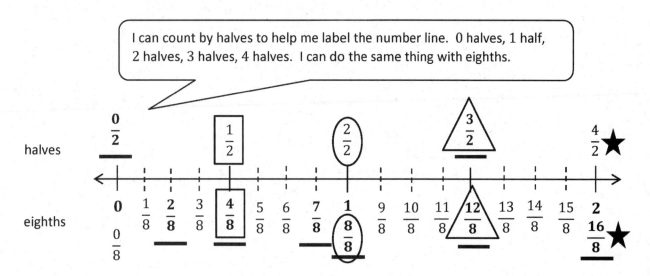

I can count by halves to help me label the number line. 0 halves, 1 half, 2 halves, 3 halves, 4 halves. I can do the same thing with eighths.

2. Use the number line above to do the following:

   ▪ Circle fractions equal to 1.

   ▪ Draw a box around fractions equal to 1 half.

   ▪ Draw a star next to fractions equal to 2.

   ▪ Draw a triangle around fractions equal to 3 halves.

   ▪ Write a pair of fractions that are equivalent.

   I know that equivalent fractions are at the same point on the number line. I can see that $\frac{2}{2}$ and $\frac{8}{8}$ are equal to 1 because they are at the same point on the number line.

$$\frac{\dfrac{3}{2}}{\rule{2cm}{0.4pt}} = \frac{\dfrac{12}{8}}{\rule{2cm}{0.4pt}}$$

   $\frac{3}{2}$ and $\frac{12}{8}$ are equivalent fractions because they are at the same point on the number line.

Lesson 21: Recognize and show that equivalent fractions refer to the same point on the number line.

89

EUREKA MATH®

Name _____ Date _____

1. Use the fractional units on the left to count up on the number line. Label the missing fractions on the blanks.

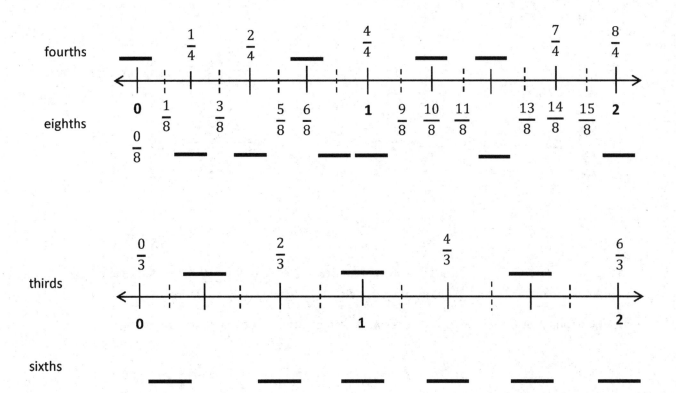

2. Use the number lines above to:

   • Color fractions equal to 1 purple.

   • Color fractions equal to 2 fourths yellow.

   • Color fractions equal to 2 blue.

   • Color fractions equal to 5 thirds green.

   • Write a pair of fractions that are equivalent.

   _____ = _____

Lesson 21: Recognize and show that equivalent fractions refer to the same point on the number line.

© 2018 Great Minds®. eureka-math.org

91

3. Use the number lines on the previous page to make the number sentences true.

$$\frac{1}{4} = \frac{}{8} \qquad \frac{6}{4} = \frac{12}{} \qquad \frac{2}{3} = \frac{}{6}$$

$$\frac{6}{3} = \frac{12}{} \qquad \frac{3}{3} = \frac{}{6} \qquad 2 = \frac{8}{4} = \frac{}{8}$$

4. Mr. Fairfax ordered 3 large pizzas for a class party. Group A ate $\frac{6}{6}$ of the first pizza, and Group B ate $\frac{8}{6}$ of the remaining pizza. During the party, the class discussed which group ate more pizza.

   a. Did Group A or B eat more pizza? Use words and pictures to explain your answer to the class.

   b. Later, Group C ate all remaining slices of pizza. What fraction of the pizza did group C eat? Use words and pictures to explain your answer.

Lesson 21: Recognize and show that equivalent fractions refer to the same point on the number line.

© 2018 Great Minds®. eureka-math.org

1. Write the shaded fraction of each figure on the blank. Then, draw a line to match the equivalent fractions.

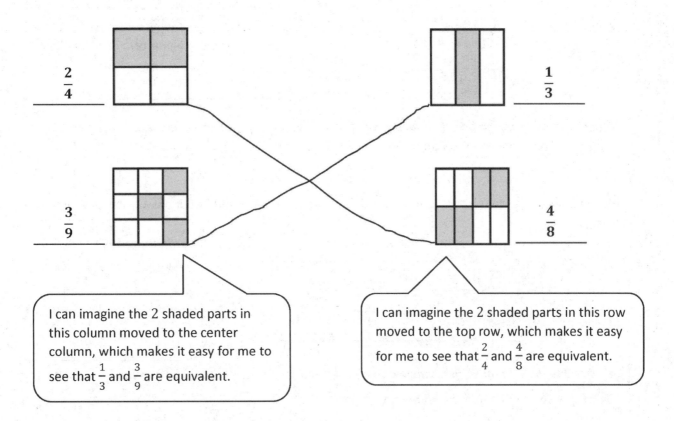

$\dfrac{2}{4}$

$\dfrac{1}{3}$

$\dfrac{3}{9}$

$\dfrac{4}{8}$

I can imagine the 2 shaded parts in this column moved to the center column, which makes it easy for me to see that $\dfrac{1}{3}$ and $\dfrac{3}{9}$ are equivalent.

I can imagine the 2 shaded parts in this row moved to the top row, which makes it easy for me to see that $\dfrac{2}{4}$ and $\dfrac{4}{8}$ are equivalent.

2. Complete the fraction to make a true statement.

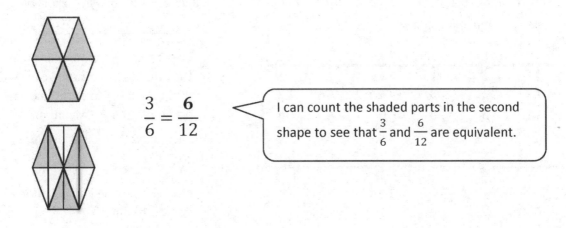

$$\dfrac{3}{6} = \dfrac{6}{12}$$

I can count the shaded parts in the second shape to see that $\dfrac{3}{6}$ and $\dfrac{6}{12}$ are equivalent.

Lesson 22:   Generate simple equivalent fractions by using visual fraction models and the number line.

93

© 2018 Great Minds®. eureka-math.org

3. Why does it take 2 copies of $\frac{1}{4}$ to show the same amount as 1 copy of $\frac{1}{2}$? Explain your answer in words and pictures.

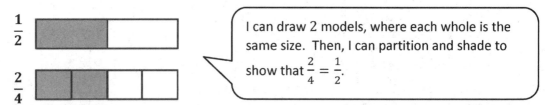

$\frac{1}{2}$

$\frac{2}{4}$

I can draw 2 models, where each whole is the same size. Then, I can partition and shade to show that $\frac{2}{4} = \frac{1}{2}$.

There is double the number of equal parts in fourths than halves, so you need double the number of copies to show equivalent fractions.

4. How many eighths does it take to make the same amount as $\frac{1}{4}$? Explain your answer in words and pictures.

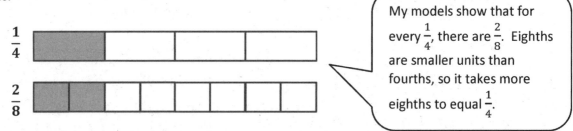

$\frac{1}{4}$

$\frac{2}{8}$

My models show that for every $\frac{1}{4}$, there are $\frac{2}{8}$. Eighths are smaller units than fourths, so it takes more eighths to equal $\frac{1}{4}$.

It takes 2 eighths to make the same amount as $\frac{1}{4}$ because there is double the number of equal parts in eighths, so it takes double the number of copies.

5. A pizza was cut into 6 equal slices. If Lizzie ate $\frac{1}{3}$ of the pizza, how many slices did she eat? Explain your answer using a number line and words.

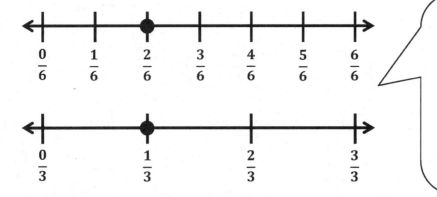

$\frac{0}{6}$ $\frac{1}{6}$ $\frac{2}{6}$ $\frac{3}{6}$ $\frac{4}{6}$ $\frac{5}{6}$ $\frac{6}{6}$

$\frac{0}{3}$ $\frac{1}{3}$ $\frac{2}{3}$ $\frac{3}{3}$

I can draw two number lines that are the same size. I can partition one into sixths and the other into thirds. My number lines show that $\frac{1}{3}$ is equivalent to $\frac{2}{6}$. I also could have drawn one number line and partition it into thirds and sixths.

Lizzie ate 2 slices of pizza because my number lines show that $\frac{1}{3} = \frac{2}{6}$, and $\frac{2}{6}$ means that she ate 2 of the 6 pieces.

Lesson 22:   Generate simple equivalent fractions by using visual fraction models and the number line.

EUREKA MATH®

Name _____   Date _____

1. Write the shaded fraction of each figure on the blank. Then, draw a line to match the equivalent fractions.

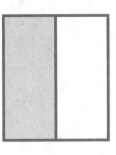

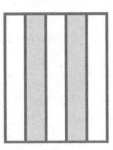

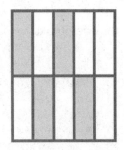

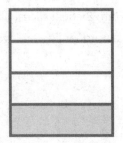

Lesson 22: Generate simple equivalent fractions by using visual fraction models and the number line.

95

© 2018 Great Minds®. eureka-math.org

2. Complete the fractions to make true statements.

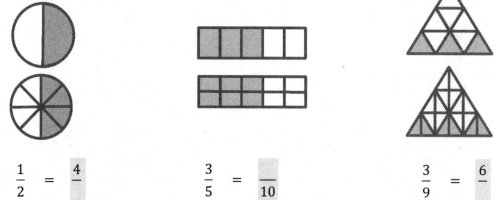

$$\frac{1}{2} = \frac{4}{}$$

$$\frac{3}{5} = \frac{}{10}$$

$$\frac{3}{9} = \frac{6}{}$$

3. Why does it take 3 copies of $\frac{1}{6}$ to show the same amount as 1 copy of $\frac{1}{2}$? Explain your answer in words and pictures.

4. How many ninths does it take to make the same amount as $\frac{1}{3}$? Explain your answer in words and pictures.

5. A pie was cut into 8 equal slices. If Ruben ate $\frac{3}{4}$ of the pie, how many slices did he eat? Explain your answer using a number line and words.

Lesson 22:   Generate simple equivalent fractions by using visual fraction models and the number line.

© 2018 Great Minds®. eureka-math.org

EUREKA MATH

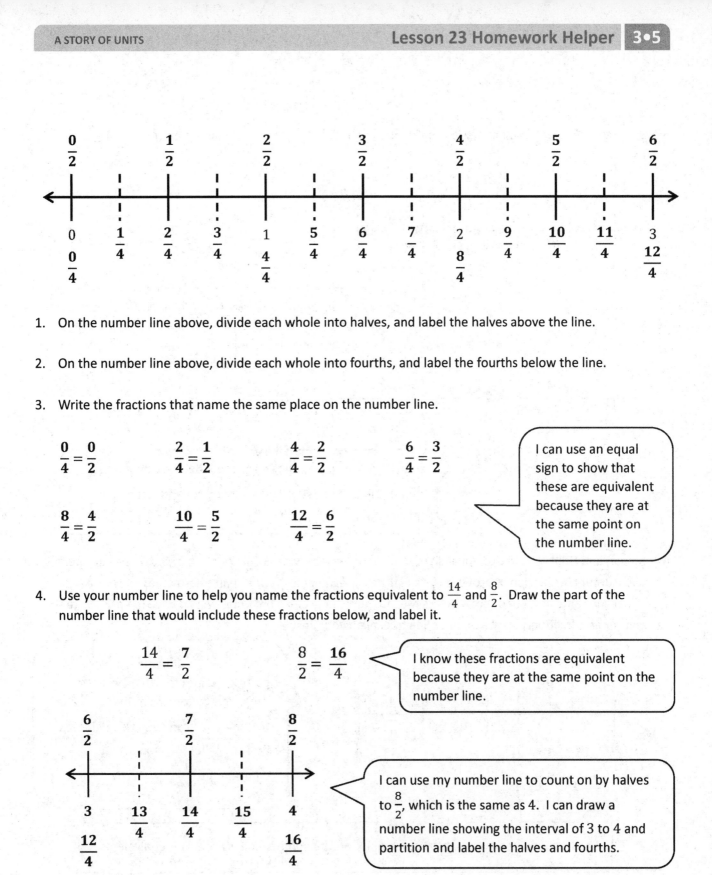

1. On the number line above, divide each whole into halves, and label the halves above the line.

2. On the number line above, divide each whole into fourths, and label the fourths below the line.

3. Write the fractions that name the same place on the number line.

$\frac{0}{4} = \frac{0}{2}$ $\frac{2}{4} = \frac{1}{2}$ $\frac{4}{4} = \frac{2}{2}$ $\frac{6}{4} = \frac{3}{2}$

$\frac{8}{4} = \frac{4}{2}$ $\frac{10}{4} = \frac{5}{2}$ $\frac{12}{4} = \frac{6}{2}$

> I can use an equal sign to show that these are equivalent because they are at the same point on the number line.

4. Use your number line to help you name the fractions equivalent to $\frac{14}{4}$ and $\frac{8}{2}$. Draw the part of the number line that would include these fractions below, and label it.

$\frac{14}{4} = \frac{7}{2}$ $\frac{8}{2} = \frac{16}{4}$

> I know these fractions are equivalent because they are at the same point on the number line.

> I can use my number line to count on by halves to $\frac{8}{2}$, which is the same as 4. I can draw a number line showing the interval of 3 to 4 and partition and label the halves and fourths.

EUREKA MATH

Lesson 23:   Generate simple equivalent fractions by using visual fraction models and the number line.

97

© 2018 Great Minds®. eureka-math.org

5. Write two different fraction names for the dot on the number line. You may use halves, fourths, or eighths.

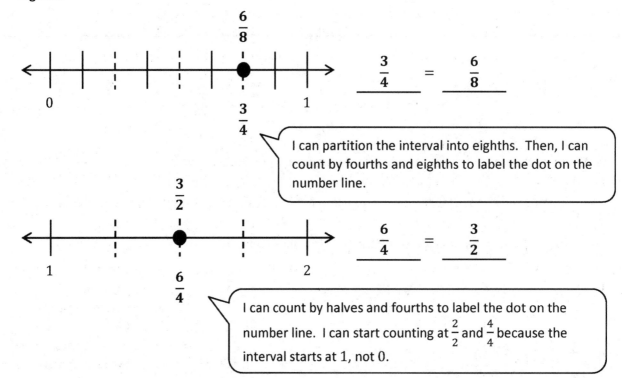

$$\frac{3}{4} = \frac{6}{8}$$

> I can partition the interval into eighths. Then, I can count by fourths and eighths to label the dot on the number line.

$$\frac{6}{4} = \frac{3}{2}$$

> I can count by halves and fourths to label the dot on the number line. I can start counting at $\frac{2}{2}$ and $\frac{4}{4}$ because the interval starts at 1, not 0.

6. Megan and Hunter bake two equal-sized pans of brownies. Megan cuts her pan of brownies into fourths, and Hunter cuts his pan of brownies into eighths. Megan eats $\frac{1}{4}$ of her pan of brownies. If Hunter wants to eat the same amount of brownies as Megan, how many of his brownies will he have to eat? Write the answer as a fraction. Draw a number line to explain your answer.

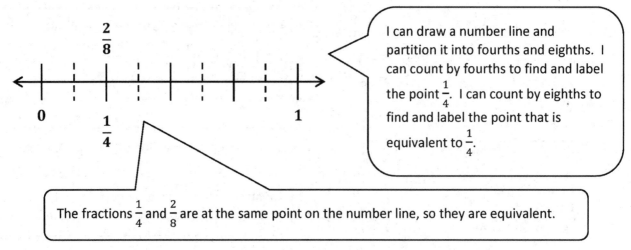

> I can draw a number line and partition it into fourths and eighths. I can count by fourths to find and label the point $\frac{1}{4}$. I can count by eighths to find and label the point that is equivalent to $\frac{1}{4}$.

> The fractions $\frac{1}{4}$ and $\frac{2}{8}$ are at the same point on the number line, so they are equivalent.

Hunter needs to eat $\frac{2}{8}$ of his brownies to eat the same amount as Megan because $\frac{2}{8} = \frac{1}{4}$.

Lesson 23:   Generate simple equivalent fractions by using visual fraction models and the number line.

Name _____   Date _____

```
   ←+———————+———————+———————+———→
    0        1        2        3
```

1. On the number line above, use a colored pencil to divide each whole into thirds and label each fraction above the line.

2. On the number line above, use a different colored pencil to divide each whole into sixths and label each fraction below the line.

3. Write the fractions that name the same place on the number line.

4. Using your number line to help, name the fraction equivalent to $\frac{20}{6}$. Name the fraction equivalent to $\frac{12}{3}$. Draw the part of the number line that would include these fractions below, and label it.

$$\frac{20}{6} = \frac{}{3}$$

$$\frac{12}{3} = \frac{}{6}$$

Lesson 23:   Generate simple equivalent fractions by using visual fraction models and the number line.

99

5. Write two different fraction names for the dot on the number line. You may use halves, thirds, fourths, fifths, sixths, eighths, or tenths.

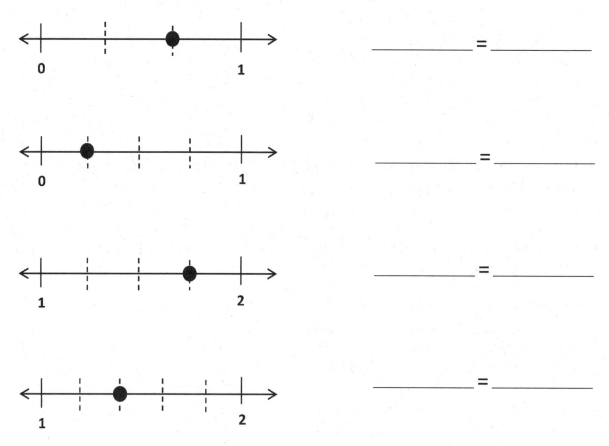

_____ = _____

_____ = _____

_____ = _____

_____ = _____

6. Danielle and Mandy each ordered a large pizza for dinner. Danielle's pizza was cut into sixths, and Mandy's pizza was cut into twelfths. Danielle ate 2 sixths of her pizza. If Mandy wants to eat the same amount of pizza as Danielle, how many slices of pizza will she have to eat? Write the answer as a fraction. Draw a number line to explain your answer.

Lesson 23:   Generate simple equivalent fractions by using visual fraction models and the number line.

1. Complete the number bond as indicated by the fractional unit. Partition the number line into the given fractional unit, and label the fractions. Rename 0 and 1 as fractions of the given unit.

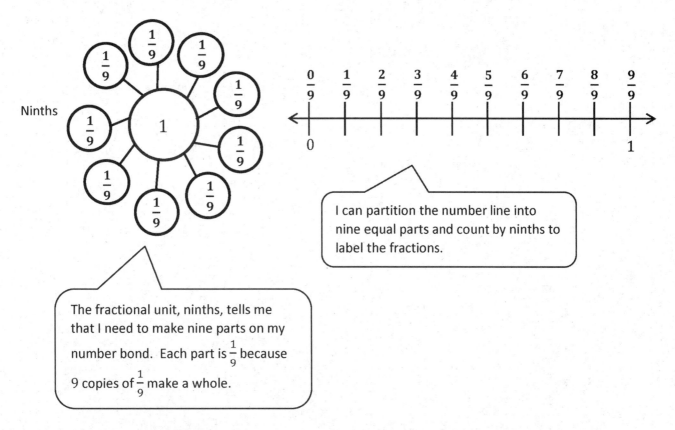

I can partition the number line into nine equal parts and count by ninths to label the fractions.

The fractional unit, ninths, tells me that I need to make nine parts on my number bond. Each part is $\frac{1}{9}$ because 9 copies of $\frac{1}{9}$ make a whole.

2. Mrs. Smith bakes two large apple pies. She cuts one pie into fourths and gives it to her daughter. She cuts the other pie into eighths and gives it to her son. Her son says, "My pie is bigger because it has more pieces than yours!" Is he correct? Use words, pictures, or a number line to help you explain.

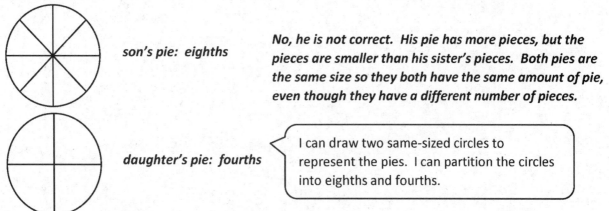

son's pie: eighths

daughter's pie: fourths

No, he is not correct. His pie has more pieces, but the pieces are smaller than his sister's pieces. Both pies are the same size so they both have the same amount of pie, even though they have a different number of pieces.

I can draw two same-sized circles to represent the pies. I can partition the circles into eighths and fourths.

Lesson 24: Express whole numbers as fractions and recognize equivalence with different units.

© 2018 Great Minds®. eureka-math.org

101

Name _____   Date _____

1. Complete the number bond as indicated by the fractional unit. Partition the number line into the given fractional unit, and label the fractions. Rename 0 and 1 as fractions of the given unit.

Fifths

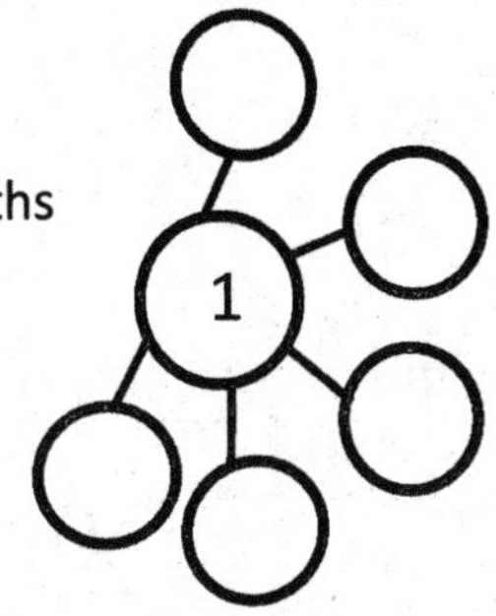

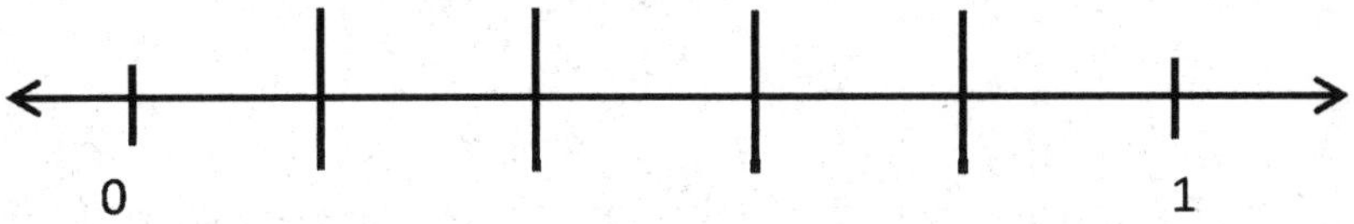

Sixths

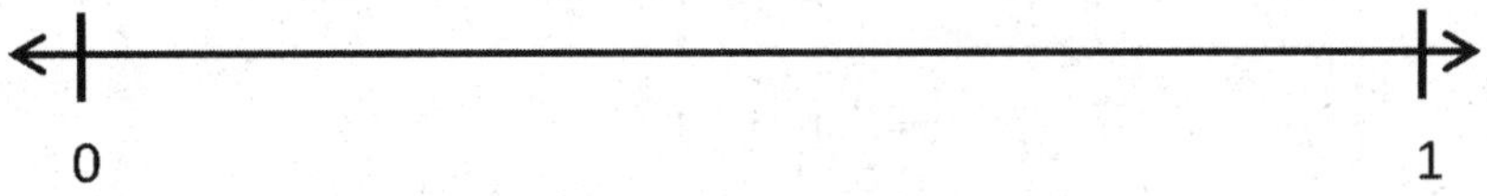

Sevenths

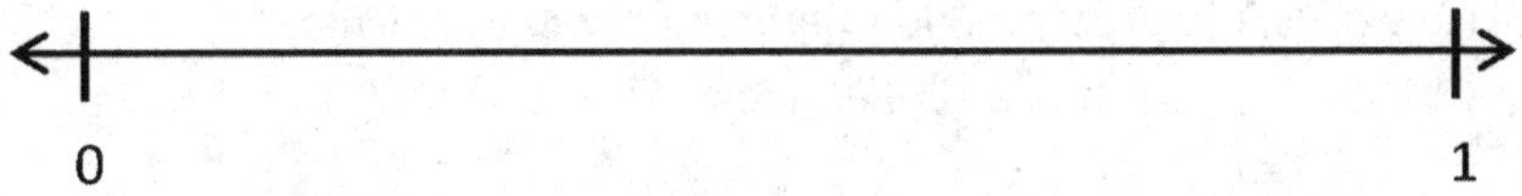

Eighths

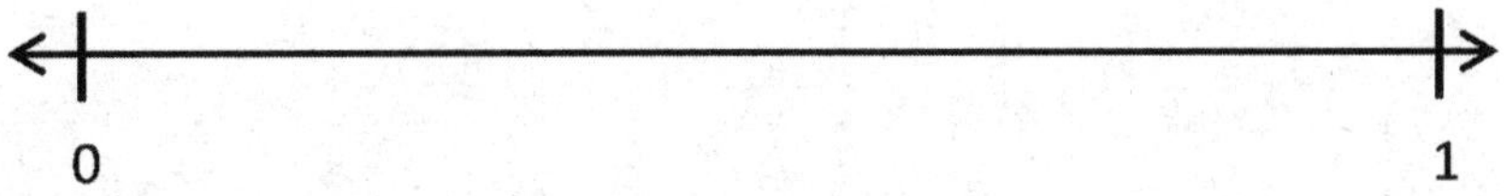

Lesson 24:   Express whole numbers as fractions and recognize equivalence with different units.

© 2018 Great Minds®. eureka-math.org

103

EUREKA MATH

2. Circle all the fractions in Problem 1 that are equal to 1. Write them in a number sentence below.

$\frac{5}{5}$ = _____ = _____ = _____

3. What pattern do you notice in the fractions that are equivalent to 1? Following this pattern, how would you represent ninths as 1 whole?

4. In Art class, Mr. Joselyn gave everyone a 1-foot stick to measure and cut. Vivian measured and cut her stick into 5 equal pieces. Scott measured and cut his into 7 equal pieces. Scott said to Vivian, "The total length of my stick is longer than yours because I have 7 pieces, and you only have 5." Is Scott correct? Use words, pictures, or a number line to help you explain.

Lesson 24:   Express whole numbers as fractions and recognize equivalence with different units.

© 2018 Great Minds®. eureka-math.org

EUREKA MATH

1. Label the following models as fractions inside the boxes.

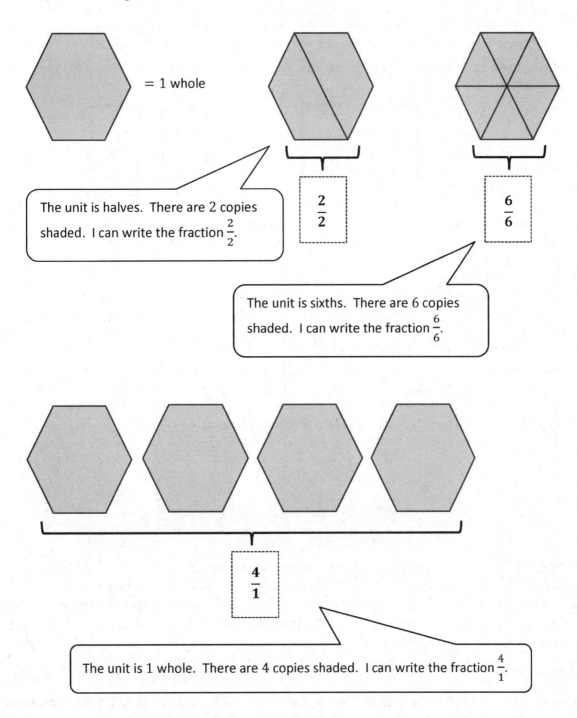

= 1 whole

$\dfrac{2}{2}$

$\dfrac{6}{6}$

The unit is halves. There are 2 copies shaded. I can write the fraction $\dfrac{2}{2}$.

The unit is sixths. There are 6 copies shaded. I can write the fraction $\dfrac{6}{6}$.

$\dfrac{4}{1}$

The unit is 1 whole. There are 4 copies shaded. I can write the fraction $\dfrac{4}{1}$.

EUREKA MATH®

Lesson 25:   Express whole number fractions on the number line when the unit interval is 1.

© 2018 Great Minds®. eureka-math.org

105

2. Fill in the missing whole numbers in the boxes below the number line. Use the pattern to rename the whole numbers as fractions in the boxes above the number line.

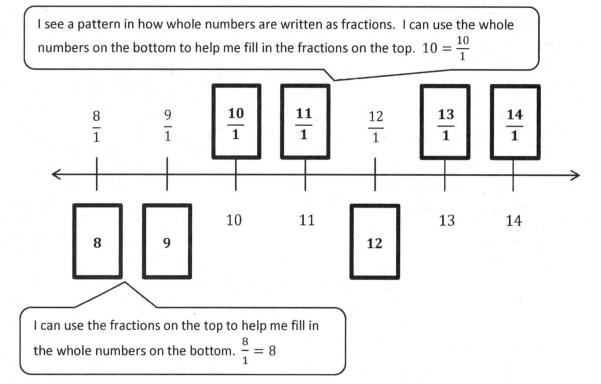

I see a pattern in how whole numbers are written as fractions. I can use the whole numbers on the bottom to help me fill in the fractions on the top. $10 = \frac{10}{1}$

$\frac{8}{1}$ $\frac{9}{1}$ $\boxed{\frac{10}{1}}$ $\boxed{\frac{11}{1}}$ $\frac{12}{1}$ $\boxed{\frac{13}{1}}$ $\boxed{\frac{14}{1}}$

$\boxed{8}$ $\boxed{9}$ 10   11 $\boxed{12}$ 13   14

I can use the fractions on the top to help me fill in the whole numbers on the bottom. $\frac{8}{1} = 8$

3. Explain the difference between these fractions with words and pictures.

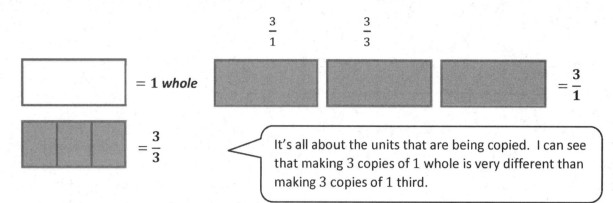

$\frac{3}{1}$ $\frac{3}{3}$

$\boxed{}$ = 1 whole

$= \frac{3}{1}$

$= \frac{3}{3}$

It's all about the units that are being copied. I can see that making 3 copies of 1 whole is very different than making 3 copies of 1 third.

The fractions $\frac{3}{1}$ and $\frac{3}{3}$ are different because they both represent 3 copies, but the units that are copied are different. The fraction $\frac{3}{1}$ is 3 copies of 1 whole, and the fraction $\frac{3}{3}$ is 3 copies of 1 third. 3 copies of 1 whole, or $\frac{3}{1}$, is greater than 3 copies of 1 third, or $\frac{3}{3}$. My picture shows that $\frac{3}{1}$ is 3 wholes, and $\frac{3}{3}$ is only 1 whole.

Lesson 25:   Express whole number fractions on the number line when the unit interval is 1.

EUREKA MATH

Name _____   Date _____

1. Label the following models as fractions inside the boxes.

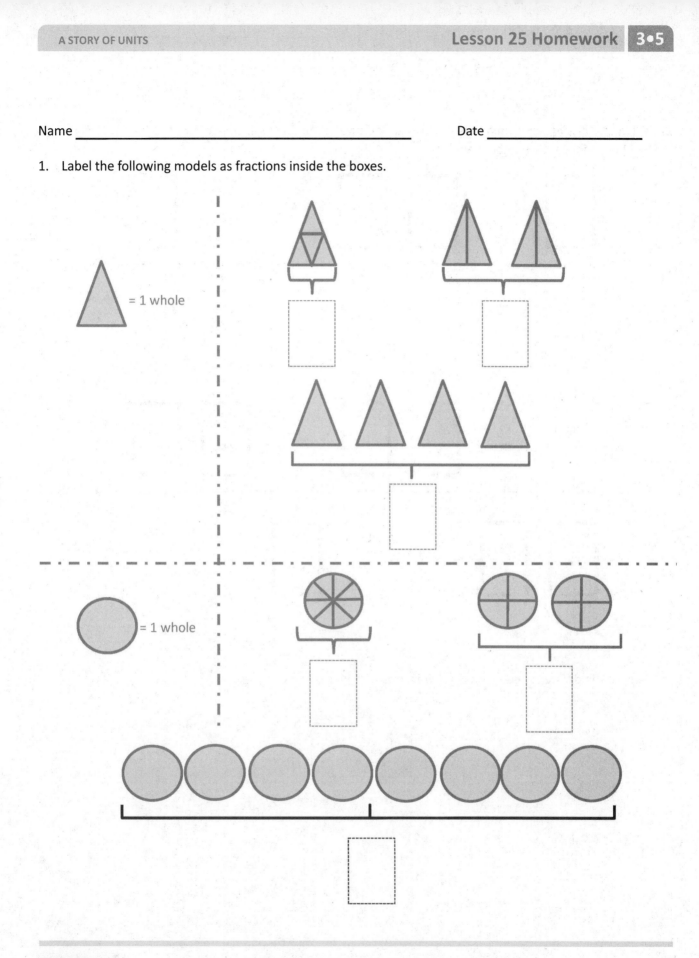

Lesson 25:   Express whole number fractions on the number line when the unit interval is 1.

© 2018 Great Minds®. eureka-math.org

107

2. Fill in the missing whole numbers in the boxes below the number line. Rename the wholes as fractions in the boxes above the number line.

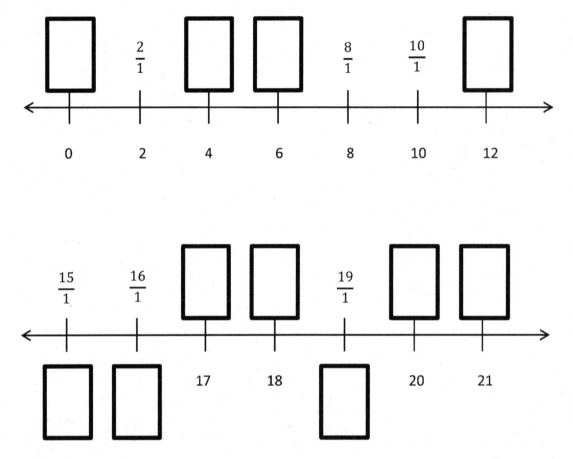

3. Explain the difference between these fractions with words and pictures.

$$\frac{5}{1} \qquad \frac{5}{5}$$

Lesson 25:   Express whole number fractions on the number line when the unit interval is 1.

© 2018 Great Minds®. eureka-math.org

1. Partition the number line to show the fractional units. Then, draw number bonds with copies of 1 whole for the circled whole numbers.

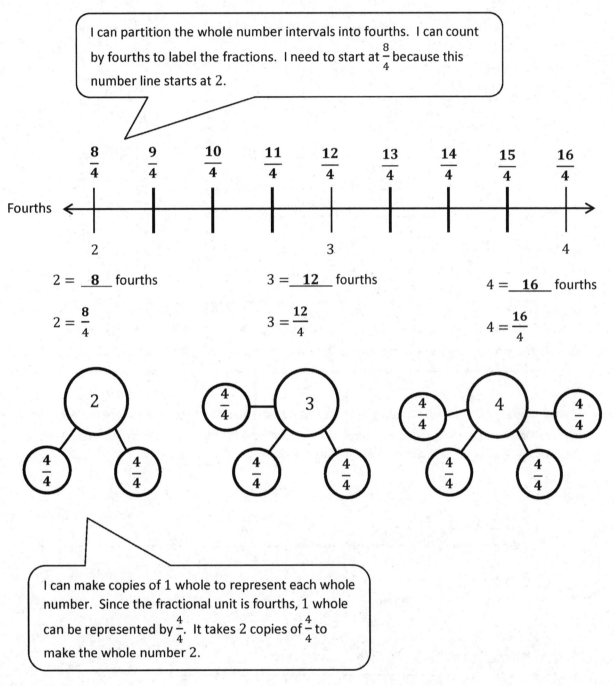

I can partition the whole number intervals into fourths. I can count by fourths to label the fractions. I need to start at $\frac{8}{4}$ because this number line starts at 2.

$2 = \underline{\textbf{8}}$ fourths

$3 = \underline{\textbf{12}}$ fourths

$4 = \underline{\textbf{16}}$ fourths

$2 = \frac{8}{4}$

$3 = \frac{12}{4}$

$4 = \frac{16}{4}$

I can make copies of 1 whole to represent each whole number. Since the fractional unit is fourths, 1 whole can be represented by $\frac{4}{4}$. It takes 2 copies of $\frac{4}{4}$ to make the whole number 2.

Lesson 26: Decompose whole number fractions greater than 1 using whole number equivalence with various models.

© 2018 Great Minds®. eureka-math.org

109

2. Use the number line to write the fractions that name the whole numbers for each fractional unit. The first one has been done for you.

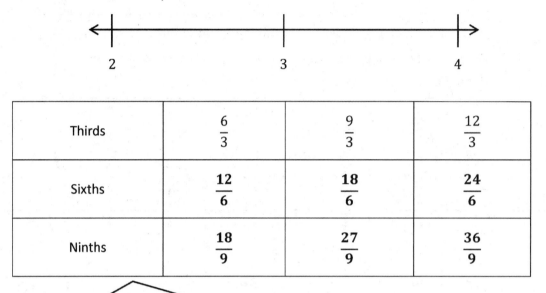

Thirds	$\frac{6}{3}$	$\frac{9}{3}$	$\frac{12}{3}$
Sixths	$\frac{12}{6}$	$\frac{18}{6}$	$\frac{24}{6}$
Ninths	$\frac{18}{9}$	$\frac{27}{9}$	$\frac{36}{9}$

I know that $\frac{12}{6} = 2$. I can count by sixths to find the other fractions that name the whole numbers on the number line. I can do the same thing for ninths.

3. Monica walks $\frac{1}{4}$ of a mile on Monday. Each day after that, she walks $\frac{1}{4}$ of a mile more than she did the day before. Draw and partition a number line to represent how far Monica walks on Monday, Tuesday, Wednesday, and Thursday. What fraction of a mile does she walk on Thursday?

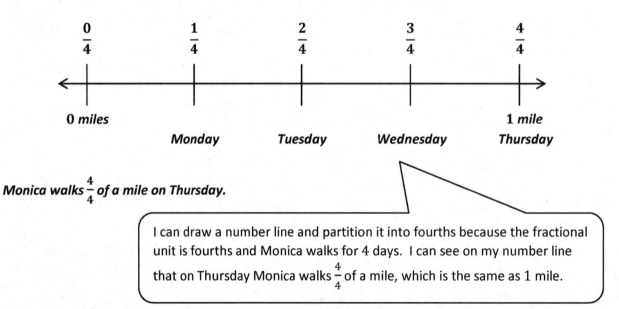

Monica walks $\frac{4}{4}$ of a mile on Thursday.

I can draw a number line and partition it into fourths because the fractional unit is fourths and Monica walks for 4 days. I can see on my number line that on Thursday Monica walks $\frac{4}{4}$ of a mile, which is the same as 1 mile.

Lesson 26: Decompose whole number fractions greater than 1 using whole number equivalence with various models.

EUREKA MATH

Name _____ Date _____

1. Partition the number line to show the fractional units. Then, draw number bonds with copies of 1 whole for the circled whole numbers.

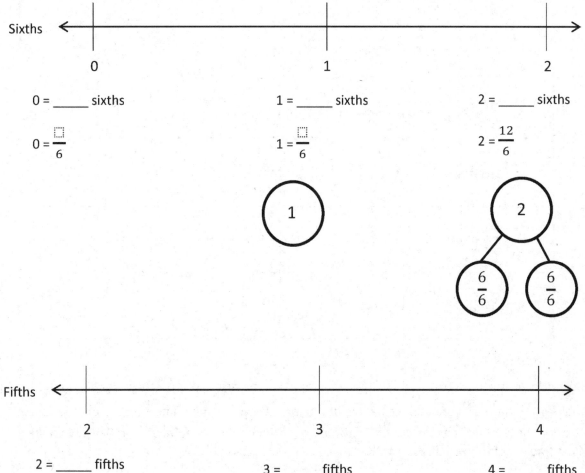

Sixths

0   1   2

0 = _____ sixths   1 = _____ sixths   2 = _____ sixths

$0 = \dfrac{\square}{6}$ $1 = \dfrac{\square}{6}$ $2 = \dfrac{12}{6}$

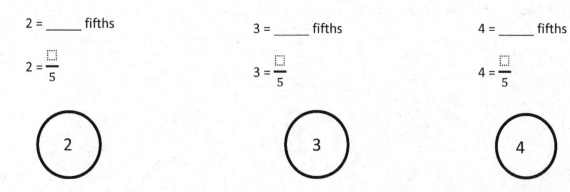

Fifths

2   3   4

2 = _____ fifths   3 = _____ fifths   4 = _____ fifths

$2 = \dfrac{\square}{5}$ $3 = \dfrac{\square}{5}$ $4 = \dfrac{\square}{5}$

Lesson 26:   Decompose whole number fractions greater than 1 using whole
             number equivalence with various models.

© 2018 Great Minds®. eureka-math.org

111

2. Write the fractions that name the whole numbers for each fractional unit. The first one has been done for you.

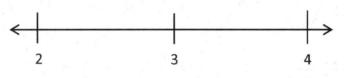

	2	3	4
Thirds	$\frac{6}{3}$	$\frac{9}{3}$	$\frac{12}{3}$
Sevenths			
Eighths			
Tenths			

3. Rider dribbles the ball down $\frac{1}{3}$ of the basketball court on the first day of practice. Each day after that, he dribbles $\frac{1}{3}$ of the way more than he did the day before. Draw a number line to represent the court. Partition the number line to represent how far Rider dribbles on Day 1, Day 2, and Day 3 of practice. What fraction of the way does he dribble on Day 3?

Lesson 26:   Decompose whole number fractions greater than 1 using whole number equivalence with various models.

© 2018 Great Minds®. eureka-math.org

1. Use the pictures to model equivalent fractions. Fill in the blanks, and answer the questions.

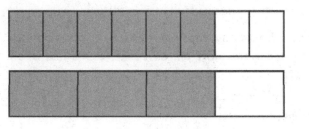

> I can shade 6 eighths, and then I can shade fourths until the same amount in each model is shaded. It takes 3 fourths to equal 6 eighths.

6 eighths is equal to __3__ fourths.

$$\frac{6}{8} = \frac{3}{4}$$

The whole stays the same.

What happens to the size of the equal parts when there are fewer equal parts?

When there are fewer equal parts, the size of each equal part gets bigger. Fourths are bigger than eighths.

2. Six friends share 2 crackers that are both the same size. The crackers are represented by the 2 rectangles below. The first cracker is cut into 3 equal parts, and the second is cut into 6 equal parts. How can the 6 friends share the crackers equally without breaking any of the pieces?

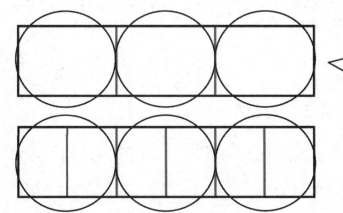

> I can partition the first cracker into thirds and the second cracker into sixths. I can circle 6 equal amounts to show how much each friend gets.

Three friends each get $\frac{1}{3}$ of the first cracker. The other 3 friends each get $\frac{2}{6}$ of the second cracker. They all get the same amount because $\frac{1}{3} = \frac{2}{6}$.

EUREKA MATH®

Lesson 27:   Explain equivalence by manipulating units and reasoning about their size.

© 2018 Great Minds®. eureka-math.org

113

3. Mrs. Mills cuts a pizza into 6 equal slices. Then, she cuts every slice in half. How many of the smaller slices does she have? Use words and numbers to explain your answer.

She has 12 smaller slices of pizza. Since she cut each slice in half, that means that she doubled the number of pieces and $6 \times 2 = 12$. The smaller the pieces, the more pieces it takes to make a whole.

If I need to, I can draw a picture. I can draw a circle and partition it into sixths. Then, I can partition each sixth into 2 equal pieces. That would make 12 pieces.

Lesson 27:   Explain equivalence by manipulating units and reasoning about their size.

Name _____   Date _____

1. Use the pictures to model equivalent fractions. Fill in the blanks, and answer the questions.

2 tenths is equal to _____ fifths.

$$\frac{2}{10} = \frac{}{5}$$

The whole stays the same.

What happened to the size of the equal parts when there were fewer equal parts?

1 third is equal to _____ ninths.

$$\frac{1}{3} = \frac{}{9}$$

The whole stays the same.

What happened to the size of the equal parts when there were more equal parts?

2. 8 students share 2 pizzas that are the same size, which are represented by the 2 circles below. They notice that the first pizza is cut into 4 equal slices, and the second is cut into 8 equal slices. How can the 8 students share the pizzas equally without cutting any of the pieces?

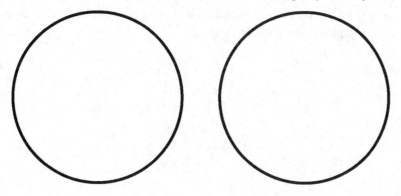

3. When the whole is the same, why does it take 4 copies of 1 tenth to equal 2 copies of 1 fifth? Draw a model to support your answer.

4. When the whole is the same, how many eighths does it take to equal 1 fourth? Draw a model to support your answer.

5. Mr. Pham cuts a cake into 8 equal slices. Then, he cuts every slice in half. How many of the smaller slices does he have? Use words and numbers to explain your answer.

Lesson 27:   Explain equivalence by manipulating units and reasoning about their size.

© 2018 Great Minds®. eureka-math.org

1. Shade the models to compare the fractions.

2 fourths

2 eighths

Which is larger, 2 fourths or 2 eighths? Why? Use words to explain.

2 *fourths* is larger than 2 *eighths* because the more times you cut the whole, the smaller the pieces get. The number of pieces I shaded is the same, but the sizes of the pieces are different. Eighths are much smaller than fourths.

2. After baseball practice, Steven and Eric each buy a 1-liter bottle of water. Steven drinks 3 sixths of his water. Eric drinks 3 fourths of his water. Who drinks more water?   Draw a picture to support your answer.

Steven: 3 sixths

Eric: 3 fourths

Eric drinks more water.

I can see from my picture that 3 fourths is greater than 3 sixths. I shaded the same number of parts, but the wholes are partitioned into different fractional units. Sixths are smaller than fourths.

Steven and Eric each buy a 1-liter bottle of water, so I need to draw my 2 wholes exactly the same size. If the size of the whole changes, I won't be able to accurately compare the 2 fractions.

Lesson 28:   Compare fractions with the same numerator pictorially.   **117**

Name _____   Date _____

Shade the models to compare the fractions. Circle the larger fraction for each problem.

1. 1 half

    1 fifth

2. 2 sevenths

    2 fourths

3. 4 fifths

    4 ninths

4. 5 sevenths

    5 tenths

5. 4 sixths

    4 fourths

6. Saleem and Edwin use inch rulers to measure the lengths of their caterpillars. Saleem's caterpillar measures 3 fourths of an inch. Edwin's caterpillar measures 3 eighths of an inch. Whose caterpillar is longer? Draw a picture to support your answer.

7. Lily and Jasmine each bake the same-sized chocolate cake. Lily puts $\frac{5}{10}$ of a cup of sugar into her cake. Jasmine puts $\frac{5}{6}$ of a cup of sugar into her cake. Who uses less sugar? Draw a picture to support your answer.

1. Draw your own model to compare the following fractions. Then, complete the number sentence by writing >, <, or =.

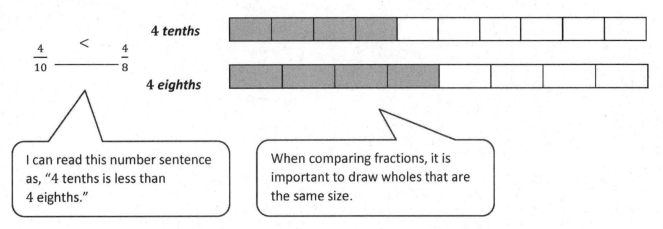

$\frac{4}{10}$ < $\frac{4}{8}$

4 tenths

4 eighths

I can read this number sentence as, "4 tenths is less than 4 eighths."

When comparing fractions, it is important to draw wholes that are the same size.

2. Draw 2 number lines with endpoints 0 and 1 to show each fraction in Problem 1. Use the number lines to explain how you know your comparison in Problem 1 is correct.

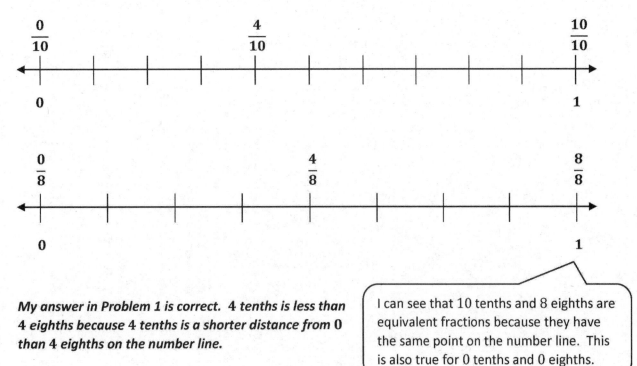

$\frac{0}{10}$ $\frac{4}{10}$ $\frac{10}{10}$

0   1

$\frac{0}{8}$ $\frac{4}{8}$ $\frac{8}{8}$

0   1

My answer in Problem 1 is correct. 4 tenths is less than 4 eighths because 4 tenths is a shorter distance from 0 than 4 eighths on the number line.

I can see that 10 tenths and 8 eighths are equivalent fractions because they have the same point on the number line. This is also true for 0 tenths and 0 eighths.

Lesson 29:   Compare fractions with the same numerator using <, >, or =, and use a model to reason about their size.

© 2018 Great Minds®. eureka-math.org

121

Name _____   Date _____

Label each shaded fraction. Use >, < or = to compare.

1.

2.

3.

4.

5. Partition each number line into the units labeled on the left. Then, use the number lines to compare the fractions.

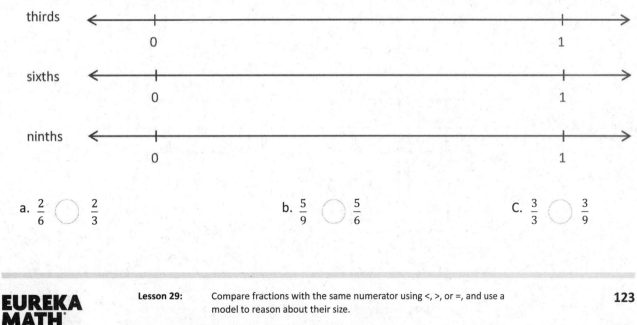

thirds

0   1

sixths

0   1

ninths

0   1

a. $\frac{2}{6}$ ◯ $\frac{2}{3}$ b. $\frac{5}{9}$ ◯ $\frac{5}{6}$ C. $\frac{3}{3}$ ◯ $\frac{3}{9}$

EUREKA
MATH

Lesson 29:   Compare fractions with the same numerator using <, >, or =, and use a
              model to reason about their size.

© 2018 Great Minds®. eureka-math.org

123

Draw your own models to compare the following fractions.

6. $\frac{7}{10}$ ◯ $\frac{7}{8}$

7. $\frac{4}{6}$ ◯ $\frac{4}{9}$

8. For an art project, Michello used $\frac{3}{4}$ of a glue stick. Yamin used $\frac{3}{6}$ of an identical glue stick. Who used more of the glue stick? Use the model below to support your answer. Be sure to label 1 whole as 1 glue stick.

9. After gym class, Jahsir drank 2 eighths of a bottle of water. Jade drank 2 fifths of an identical bottle of water. Who drank less water? Use the model below to support your answer.

Lesson 29:   Compare fractions with the same numerator using <, >, or =, and use a model to reason about their size.

© 2018 Great Minds®. eureka-math.org

Theodore precisely partitions his red strip into fifths using the number line method below. Describe step by step how Theodore partitions his strip into equal units using only a piece of notebook paper and a straight edge.

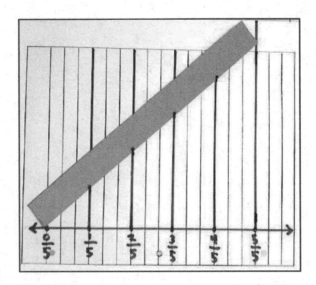

First, Theodore uses the paper's margin line to draw a number line. He then labels fifths on his number line from 0 to 1. He uses 3 spaces for each fifth. Next, at each fifth, he draws vertical lines up from the number line to the top of the paper. He then takes his red strip and angles it so that the left end touches the 0 endpoint on the number line, and the right end touches the line at 5 fifths, or 1. Finally, he marks on the red strip where the vertical points touch it. This creates equal units on the red strip. Theodore can double check by measuring them with a ruler.

Using this method, I can make fractional units precisely without a ruler. If I want to partition longer strips, like a meter strip, I tape more lined papers above the first one so that I can make a sharper angle with the longer strip.

Lesson 30:   Partition various wholes precisely into equal parts using a number line
              method.

© 2018 Great Minds®. eureka-math.org

125

Name _____ Date _____

Describe step by step the experience you had of partitioning a length into equal units by simply using a piece of notebook paper and a straight edge. Illustrate the process.

Lesson 30: Partition various wholes precisely into equal parts using a number line method.

© 2018 Great Minds®. eureka-math.org

127

Grade 3
Module 6

1. The tally chart below shows a survey of students' favorite ice cream flavors. Each tally mark represents 1 student.

Favorite Ice Cream Flavors	
Flavors	Number of Students
Chocolate	⦶⦶⦶⦶ /
Vanilla	⦶⦶⦶⦶
Cookie Dough	⦶⦶⦶⦶ //
Mint Chocolate Chip	////

> I can count the tally marks by fives and ones to find the total number of students.

The chart shows a total of _____22_____ students.

2. Use the tally chart in Problem 1 to complete the picture graph below.

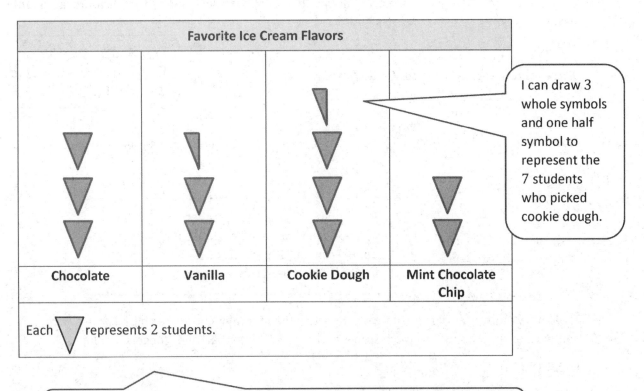

> I can draw 3 whole symbols and one half symbol to represent the 7 students who picked cookie dough.

Each ▽ represents 2 students.

> I can use the key to tell me what each symbol represents. Since each symbol represents 2 students, I can draw half a symbol to represent 1 student.

a.   What does each ▽ represent?

> I can look at the key in the picture graph to find this information.

Each ▼ represents 2 students.

b.   How many students picked vanilla as their favorite ice cream flavor?

Five students picked vanilla as their favorite ice cream flavor.

> I can look at the picture graph or the tally chart to figure out how many students picked vanilla. The picture graph shows 2 whole symbols and a half symbol, so that's 5 students.

c.   How many more students chose cookie dough than mint chocolate chip as their favorite ice cream flavor?

$7 - 4 = 3$

Three more students chose cookie dough than mint chocolate chip.

> I can find the total for each flavor and subtract to find the difference.

d.   How many students does ▽▽▽ represent? Write a number sentence to show how you know.

$3 \times 2 = 6$

$6 + 1 = 7$

It represents 7 students.

> I can multiply 3×2 because there are 3 whole symbols, and each symbol stands for 2 students. Then, I can add 1 more because there is a half symbol, which represents 1 student.

e.   How many more ▽ did you draw for chocolate than for mint chocolate chip? Write a number sentence to show how many more students chose chocolate than mint chocolate chip.

$6 - 4 = 2$

I drew 1 more symbol for chocolate than for mint chocolate chip.

> I can subtract to find the difference between the number of students who picked each flavor. The difference is 2 students. Since each symbol represents 2 students, that means I drew 1 more symbol for chocolate than for mint chocolate chip. I could also find the answer by looking at the chart and recognizing that 3 symbols for chocolate is 1 more than the 2 symbols I drew for mint chocolate chip.

Lesson 1:   Generate and organize data.

Name _____ Date _____

1. The tally chart below shows a survey of students' favorite pets. Each tally mark represents 1 student.

Favorite Pets	
Pets	**Number of Pets**
Cats	//// /
Turtles	////
Fish	//
Dogs	//// ///
Lizards	//

The chart shows a total of _____ students.

2. Use the tally chart in Problem 1 to complete the picture graph below. The first one has been done for you.

Favorite Pets				
◯ ◯ ◯ ◯ ◯ ◯				
Cats	**Turtles**	**Fish**	**Dogs**	**Lizards**

Each ◯ represents 1 student.

a. The same number of students picked _____ and _____ as their favorite pet.

b. How many students picked dogs as their favorite pet?

c. How many more students chose cats than turtles as their favorite pet?

3. Use the tally chart in Problem 1 to complete the picture graph below.

Favorite Pets				
Cats	**Turtles**	**Fish**	**Dogs**	**Lizards**

Each ☐ represents 2 students.

a. What does each ☐ represent?

b. How many students does ☐☐☐☐☐ represent? Write a number sentence to show how you know.

c. How many more ☐ did you draw for dogs than for fish? Write a number sentence to show how many more students chose dogs than fish.

Lesson 1:   Generate and organize data.

1. Lenny surveys third graders to find out their favorite recess activities. The results are in the table below.

Favorite Recess Activities	
Recess Activity	Number of Student Votes
Swinging	6
Tag	10
Basketball	14
Kickball	8

Draw units of 2 to complete the tape diagrams to show the total votes for each recess activity. The first one has been done for you.

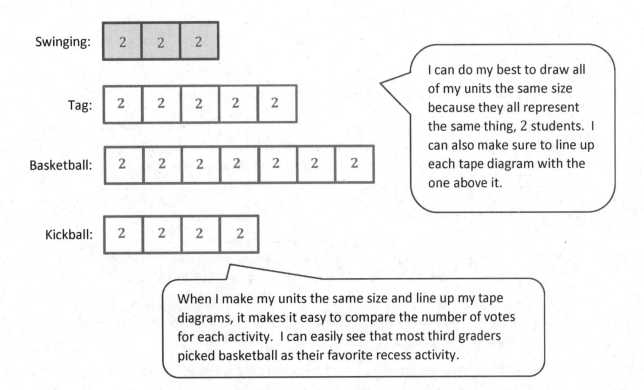

Swinging: | 2 | 2 | 2 |

Tag: | 2 | 2 | 2 | 2 | 2 |

Basketball: | 2 | 2 | 2 | 2 | 2 | 2 | 2 |

Kickball: | 2 | 2 | 2 | 2 |

I can do my best to draw all of my units the same size because they all represent the same thing, 2 students. I can also make sure to line up each tape diagram with the one above it.

When I make my units the same size and line up my tape diagrams, it makes it easy to compare the number of votes for each activity. I can easily see that most third graders picked basketball as their favorite recess activity.

2. Complete the vertical tape diagrams below using the data from Problem 1.

a.

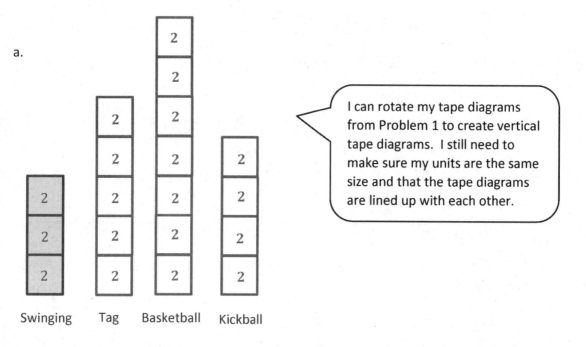

I can rotate my tape diagrams from Problem 1 to create vertical tape diagrams. I still need to make sure my units are the same size and that the tape diagrams are lined up with each other.

b. What is a good title for the vertical tape diagrams?

A good title for the vertical tape diagrams is Favorite Recess Activities.

I can use the title from the table in Problem 1 as the title for the vertical tape diagrams because they both show the same information, just in different ways.

c. Write a multiplication sentence to show the total number of votes for basketball.

$7 \times 2 = 14$

There are 7 units of 2 for basketball, so I can represent the total with the multiplication sentence $7 \times 2 = 14$.

d. If the tape diagrams in Problem 1 were made with units of 1, how would your multiplication sentence in Problem 2(c) change?

If my tape diagrams were made with units of 1 instead of 2, the multiplication sentence for Problem 2(c) would be $14 \times 1 = 14$ because there would be 14 units of 1.

Since the value of each unit is less, I need a greater number of units to represent the same total.

EUREKA MATH

Name _____   Date _____

1. Adi surveys third graders to find out their favorite fruits. The results are in the table below.

Favorite Fruits of Third Graders	
Fruit	**Number of Student Votes**
Banana	8
Apple	16
Strawberry	12
Peach	4

Draw units of 2 to complete the tape diagrams to show the total votes for each fruit. The first one has been done for you.

Banana: | 2 | 2 | 2 | 2 |

Apple:

Strawberry:

Peach:

2. Explain how you can create vertical tape diagrams to show this data.

3. Complete the vertical tape diagrams below using the data from Problem 1.

a.

b.

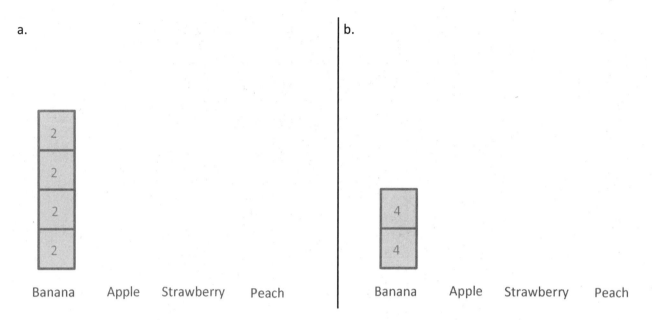

c. What is a good title for the vertical tape diagrams?

d. Compare the number of units used in the vertical tape diagrams in Problems 3(a) and 3(b). Why does the number of units change?

e. Write a multiplication number sentence to show the total number of votes for strawberry in the vertical tape diagram in Problem 3(a).

f. Write a multiplication number sentence to show the total number of votes for strawberry in the vertical tape diagram in Problem 3(b).

g. What changes in your multiplication number sentences in Problems 3(e) and (f)? Why?

Lesson 2:   Rotate tape diagrams vertically.

1. This table shows the favorite seasons of third graders.

Favorite Seasons	
Season	Number of Student Votes
Fall	16
Winter	10
Spring	13
Summer	?

Use the table to color the bar graph.

The scale on the graph tells me that each square in the grid represents 2 students. To represent the number of students who picked fall, I can color 8 squares in the grid because $8 \times 2 = 16$.

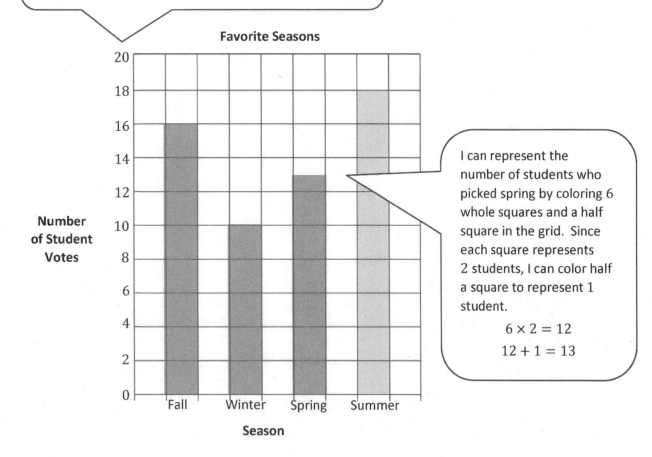

Favorite Seasons

I can represent the number of students who picked spring by coloring 6 whole squares and a half square in the grid. Since each square represents 2 students, I can color half a square to represent 1 student.

$$6 \times 2 = 12$$
$$12 + 1 = 13$$

Lesson 3:   Create scaled bar graphs.   **139**

a.   How many students voted for summer?

   18 *students voted for summer.*   ◁──── I can count by two on the bar graph to figure out how many students voted for summer.

b.   How many more students voted for fall than for spring? Write a number sentence to show your thinking.

   $16 - 13 = 3$ ◁──── I can subtract the number of students who voted for spring from the number of students who voted for fall.

   3 *more students voted for fall than for spring.*

c.   Which combination of seasons gets more votes, fall and winter together or spring and summer together? Show your work.

   Fall and winter: $16 + 10 = 26$

   Spring and summer: $13 + 18 = 31$

   The combination of spring and summer together gets more votes than fall and winter together.

   I can add the votes for fall and winter to figure out how many students voted for those two seasons. Then I can do the same thing for spring and summer. I can compare the totals to figure out which combination of seasons gets more votes.

d.   How many third graders voted in all? Show your work.

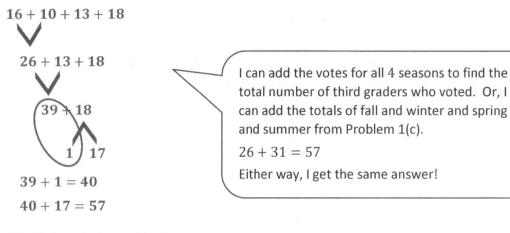

   I can add the votes for all 4 seasons to find the total number of third graders who voted. Or, I can add the totals of fall and winter and spring and summer from Problem 1(c).

   $26 + 31 = 57$

   Either way, I get the same answer!

   57 *third graders voted in all.*

Name _____ Date _____

1. This table shows the favorite subjects of third graders at Cayuga Elementary.

Favorite Subjects	
Subject	**Number of Student Votes**
Math	18
ELA	13
History	17
Science	?

Use the table to color the bar graph.

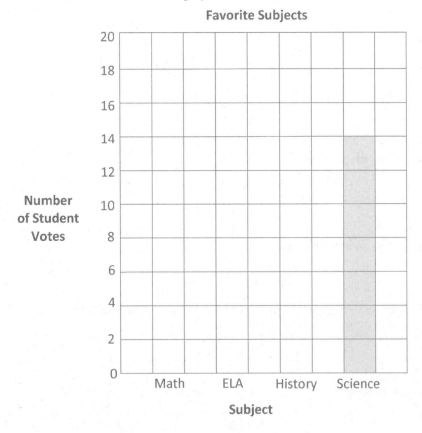

a. How many students voted for science?

b. How many more students voted for math than for science? Write a number sentence to show your thinking.

c. Which gets more votes, math and ELA together or history and science together? Show your work.

2. This bar graph shows the number of liters of water Skyler uses this month.

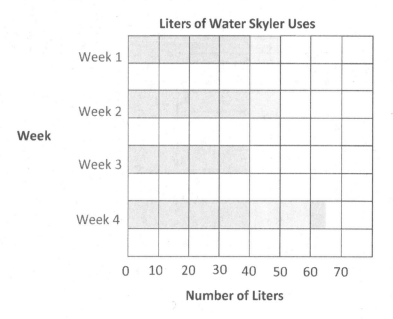

Liters of Water Skyler Uses

a. During which week does Skyler use the most water? _____
   The least? _____

b. How many more liters does Skyler use in Week 4 than Week 2?

c. Write a number sentence to show how many liters of water Skyler uses during Weeks 2 and 3 combined.

d. How many liters does Skyler use in total?

e. If Skyler uses 60 liters in each of the 4 weeks next month, will she use more or less than she uses this month? Show your work.

3. Complete the table below to show the data displayed in the bar graph in Problem 2.

Liters of Water Skyler Uses	
Week	Liters of Water

1. Farmer Brown collects the data below about the cows on his farm.

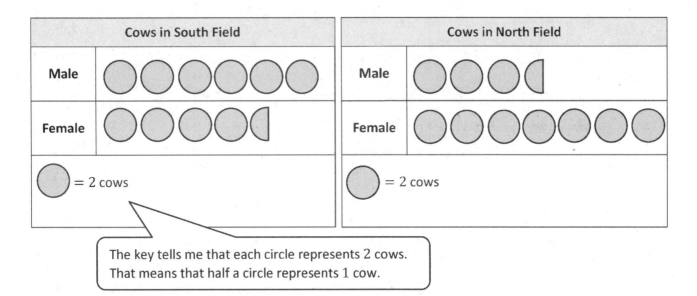

a. How many fewer male cows does Farmer Brown have than female cows?

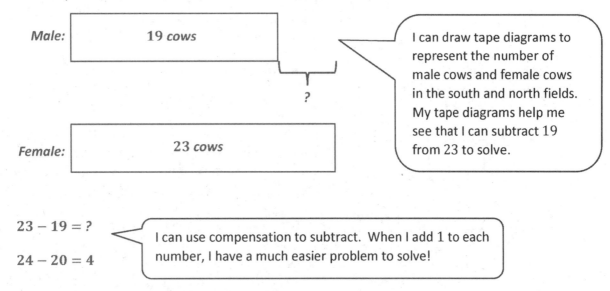

$23 - 19 = ?$

$24 - 20 = 4$

I can use compensation to subtract. When I add 1 to each number, I have a much easier problem to solve!

Farmer Brown has 4 fewer male cows than female cows.

b. It takes Farmer Brown 10 minutes to milk each female cow. How many minutes does he spend milking all of the female cows?

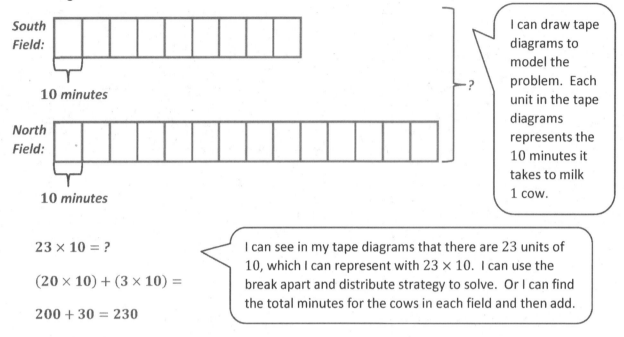

South Field:

10 *minutes*

North Field:

10 *minutes*

?

> I can draw tape diagrams to model the problem. Each unit in the tape diagrams represents the 10 minutes it takes to milk 1 cow.

$23 \times 10 = ?$

$(20 \times 10) + (3 \times 10) =$

$200 + 30 = 230$

> I can see in my tape diagrams that there are 23 units of 10, which I can represent with 23×10. I can use the break apart and distribute strategy to solve. Or I can find the total minutes for the cows in each field and then add.

Farmer Brown spends 230 *minutes milking all of the female cows.*

c. Farmer Brown's barn has 6 rows of stalls with 8 stalls in each row. How many empty stalls will there be when all the cows are in the barn?

?

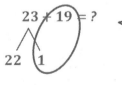

8 *stalls*

> I can draw a tape diagram to model the rows of stalls in the barn. I can multiply to find the total number of stalls.

$6 \times 8 = 48$

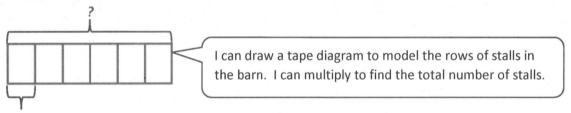

$23 + 19 = ?$

22   1

> I know there are 19 male cows and 23 female cows from my work in Problem 1(a). I can add to find the total number of cows, 42. Then, I can subtract the number of cows from the number of stalls to solve for the number of empty stalls.

$19 + 1 = 20$

$20 + 22 = 42$

$48 - 42 = 6$ *There are 6 empty stalls when all of the cows are in the barn.*

Lesson 4:   Solve one- and two-step problems involving graphs.

Name _____   Date _____

1. Maria counts the coins in her piggy bank and records the results in the tally chart below. Use the tally marks to find the total number of each coin.

Coins in Maria's Piggy Bank		
Coin	**Tally**	**Number of Coins**
Penny	///// ///// ///// ///// ///// ///// ///// ///// ///// ///// ///// ///// ///// ///// ///	
Nickel	///// ///// ///// ///// ///// ///// ///// ///// ///// ///// ///// ///// //	
Dime	///// ///// ///// ///// ///// ///// ///// ///// ///// ///// ///// //	
Quarter	///// ///// ///// ///// ////	

a. Use the tally chart to complete the bar graph below. The scale is given.

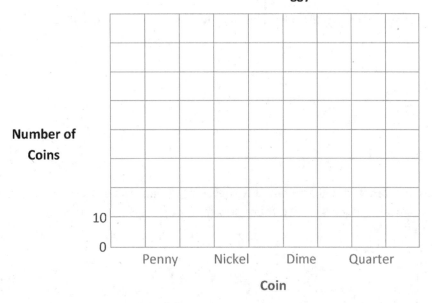

Coins in Maria's Piggy Bank

b. How many more pennies are there than dimes?

c. Maria donates 10 of each type of coin to charity. How many total coins does she have left? Show your work.

2. Ms. Hollmann's class goes on a field trip to the planetarium with Mr. Fiore's class. The number of students in each class is shown in the picture graphs below.

Students in Ms. Hollmann's Class

| Boys | ☐ ☐ ☐ ☐ ☐ ☐ |
| Girls | ☐ ☐ ☐ ☐ ☐ ☐ ☐ ☐ |

☐ = 2 students

Students in Mr. Fiore's Class

| Boys | ☐ ☐ ☐ ☐ ☐ ☐ |
| Girls | ☐ ☐ ☐ ☐ ☐ ☐ ☐ |

☐ = 2 students

a. How many fewer boys are on the trip than girls?

b. It costs $2 for each student to attend the field trip. How much money does it cost for all students to attend?

c. The cafeteria in the planetarium has 9 tables with 8 seats at each table. Counting students and teachers, how many empty seats should there be when the 2 classes eat lunch?

Lesson 4:   Solve one- and two-step problems involving graphs.

1. Samantha measures 3 crayons to the nearest inch, $\frac{1}{2}$ inch, and $\frac{1}{4}$ inch. She records the measurements in the chart below.

Crayon (color)	Measured to the Nearest Inch	Measured to the Nearest $\frac{1}{2}$ Inch	Measured to the Nearest $\frac{1}{4}$ Inch
Orange	4	$4\frac{1}{2}$	$4\frac{3}{4}$
Pink	2	$2\frac{1}{2}$	$2\frac{1}{2}$
Blue	6	6	$5\frac{3}{4}$

a. Which crayon is the longest?   *blue*

   It measures $5\frac{3}{4}$ inches.

   > The blue crayon was measured 3 times, but the most precise measurement is $5\frac{3}{4}$ inches.

b. Look carefully at Samantha's data. Which crayon most likely needs to be measured again? Explain how you know.

   The orange crayon most likely needs to be measured again. Samantha recorded 4 inches as the measurement to the nearest inch and $4\frac{3}{4}$ inches as the measurement to the nearest $\frac{1}{4}$ inch. Those measurements don't make sense. If the crayon really measures close to $4\frac{3}{4}$ inches, then the measurement to the nearest inch would be 5 inches, not 4 inches.

   > $4\frac{3}{4}$ inches is only $\frac{1}{4}$ inch away from 5 inches. It doesn't make sense for the same crayon to have measurements of $4\frac{3}{4}$ inches and 4 inches.

EUREKA MATH®

Lesson 5:   Create ruler with 1-inch, ½-inch, and ¼-inch intervals, and generate measurement data.

149

© 2018 Great Minds®. eureka-math.org

2. Evelyn marks a 3-inch paper strip into equal parts as shown below.

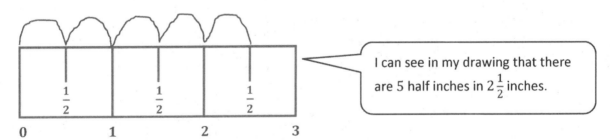

I can start at the edge of the paper strip and label it 0 inches. Then I can label the rest of the whole inches. I can label the mark halfway between each whole inch as $\frac{1}{2}$ inch.

a. Label the whole and half inches on the paper strip.

b. Estunate to draw the $\frac{1}{4}$ inch marks on the paper strip. Then, fill in the blanks below.

2 inches are equal to ___4___ half inches.

2 inches are equal to ___8___ quarter inches.

2 half inches are equal to ___4___ quarter inches.

4 quarter inches are equal to ___2___ half inches.

I can estimate to partition each $\frac{1}{2}$ inch into 2 equal parts to mark and label the $\frac{1}{4}$ inches. Then I can use the strip to help me fill in the blanks.

3. Samantha says her pink crayon measures $2\frac{1}{2}$ inches. Daniel says that's the same as 5 half inches. Explain how they are both correct.

I can see in my drawing that there are 5 half inches in $2\frac{1}{2}$ inches.

They are both correct because there are 2 half inches in each inch, so $2\frac{1}{2}$ inches is equal to 5 half inches.

Lesson 5:   Create ruler with 1-inch, ½-inch, and ¼-inch intervals, and generate
                              measurement data.

                    © 2018 Great Minds®. eureka-math.org   EUREKA MATH®

Name _____   Date _____

1. Travis measured 5 different-colored pencils to the nearest inch, $\frac{1}{2}$ inch, and $\frac{1}{4}$ inch. He records the measurements in the chart below. He draws a star next to measurements that are exact.

Colored Pencil	Measured to the nearest inch	Measured to the nearest $\frac{1}{2}$ inch	Measured to the nearest $\frac{1}{4}$ inch
Red	7	$6\frac{1}{2}$	$6\frac{3}{4}$
Blue	5	5	$5\frac{1}{4}$
Yellow	6	$5\frac{1}{2}$ ☆	$5\frac{1}{2}$ ☆
Purple	5	$4\frac{1}{2}$	$4\frac{3}{4}$
Green	2	3	$1\frac{3}{4}$

a. Which colored pencil is the longest? _____

    It measures _____ inches.

b. Look carefully at Travis's data. Which colored pencil most likely needs to be measured again? Explain how you know.

2. Evelyn marks a 4-inch paper strip into equal parts as shown below.

    a. Label the whole and half inches on the paper strip.

    b. Estimate to draw the $\frac{1}{4}$ inch marks on the paper strip. Then, fill in the blanks below.

        1 inch is equal to _____ half inches.

        1 inch is equal to _____ quarter inches

        1 half inch is equal to _____ quarter inches.

        2 quarter inches are equal to _____ half inch.

3. Travis says his yellow pencil measures $5\frac{1}{2}$ inches. Ralph says that is the same as 11 half inches. Explain how they are both correct.

Lesson 5:   Create ruler with 1-inch, ½-inch, and ¼-inch intervals, and generate
                       measurement data.

       © 2018 Great Minds®. eureka-math.org

Mr. Jackson records the amount of time his piano students spend practicing in one week. The times are shown on the line plot below.

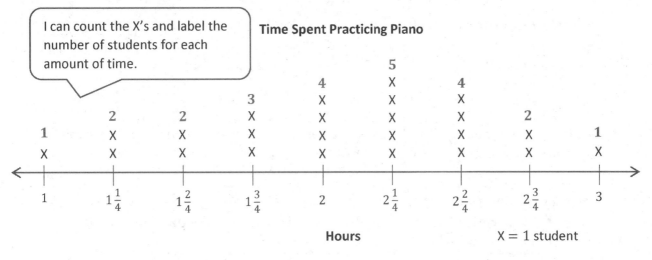

I can count the X's and label the number of students for each amount of time.

Time Spent Practicing Piano

Hours

X = 1 student

a. How many students practiced for 2 hours?

4 students practiced for 2 hours.

I can look at the labels I put on the line plot after counting to easily answer this question.

b. How many students take piano lessons from Mr. Jackson? How do you know?

24 students take lessons from Mr. Jackson. I know because I counted all of the X's on the line plot.

I can count the X's, or I can add all of the numbers that I labeled on the line plot.

$1 + 2 + 2 + 3 + 4 + 5 + 4 + 2 + 1 = 24$

c. How many students practiced for more than $2\frac{2}{4}$ hours?

3 students practiced for more than $2\frac{2}{4}$ hours.

Since it says more than $2\frac{2}{4}$ hours, I can just count the X's for $2\frac{3}{4}$ hours and 3 hours.

d. Mr. Jackson says that for students to participate in the recital, they must practice for at least 2 hours. How many students can participate in the recital?

16 *students can participate in the recital.*

> I can count the X's for the times that are equal to or more than 2 hours because the problems says, "at least 2 hours."

e. Mr. Jackson notices that the 3 most frequent times spent practicing are 2 hours, $2\frac{1}{4}$ hours, and $2\frac{2}{4}$ hours. Do you agree? Explain your answer.

Yes, I agree. 4 students practiced for both 2 hours and $2\frac{2}{4}$ hours, and 5 students practiced for $2\frac{1}{4}$ hours. These numbers of students, 4 and 5, are the most for any of the times practiced.

> I know that "most frequent times" means the times that most students spend practicing.

f. Mr. Jackson says that the most common time spent practicing is 10 quarter hours. Is he right? Why or why not?

No, he's not right. The most common time spent practicing is $2\frac{1}{4}$ hours. Since there are 4 quarter hours in each hour, there are 9 quarter hours in $2\frac{1}{4}$ hours.

$2 \times 4 = 8$

$8 + 1 = 9$

> I know that the most common time spent practicing is $2\frac{1}{4}$ hours. I find the number of quarter hours in $2\frac{1}{4}$ hours first by multiplying 2×4 because there are 2 hours, and each hour is made up of 4 quarter hours. Then I can add $8 + 1$ because there is 1 more quarter hour in the time $2\frac{1}{4}$ hours. That makes 9 quarter hours.

Name _____   Date _____

1. Ms. Leal measures the heights of the students in her kindergarten class. The heights are shown on the line plot below.

Heights of Students in Ms. Leal's Kindergarten Class

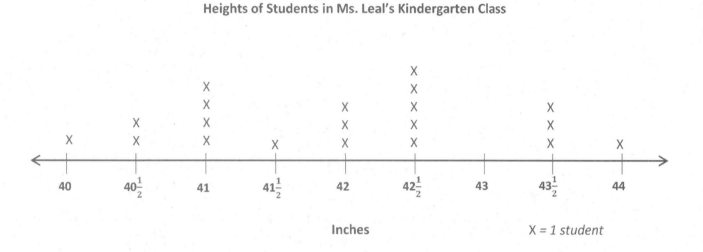

Inches   X = 1 student

a. How many students in Ms. Leal's class are exactly 41 inches tall?

b. How many students are in Ms. Leal's class? How do you know?

c. How many students in Ms. Leal's class are more than 42 inches tall?

d. Ms. Leal says that for the class picture students in the back row must be at least $42\frac{1}{2}$ inches tall. How many students should be in the back row?

2. Mr. Stein's class is studying plants. They plant seeds in clear plastic bags and measure the lengths of the roots. The lengths of the roots in inches are shown in the line plot below.

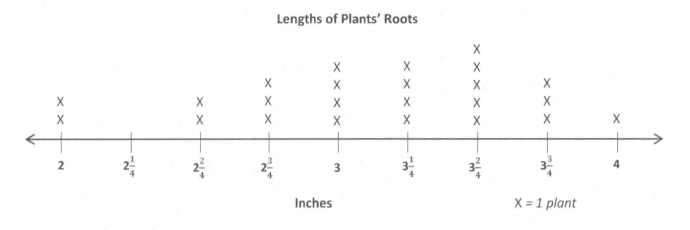

Lengths of Plants' Roots

Inches   X = 1 plant

a. How many roots did Mr. Stein's class measure? How do you know?

b. Teresa says that the 3 most frequent measurements in order from shortest to longest are $3\frac{1}{4}$ inches, $3\frac{2}{4}$ inches, and $3\frac{3}{4}$ inches. Do you agree? Explain your answer.

c. Gerald says that the most common measurement is 14 quarter inches. Is he right? Why or why not?

Lesson 6:   Interpret measurement data from various line plots.

1. The table below shows the amount of time students in Mrs. Bishop's class spent doing homework on Monday night.

Hours Spent Doing Homework				
$1\frac{1}{4}$ ✓	$\frac{3}{4}$ ✓	$\frac{1}{4}$ ✓	$\frac{1}{2}$ ✓	$1\frac{1}{2}$ ✓
$\frac{3}{4}$ ✓	1 ✓	$\frac{3}{4}$ ✓	1 ✓	$\frac{1}{2}$ ✓
0 ✓	$\frac{1}{2}$ ✓	$\frac{3}{4}$ ✓	$\frac{1}{2}$ ✓	$\frac{3}{4}$ ✓
1 ✓	$\frac{1}{4}$ ✓	$\frac{1}{4}$ ✓	1 ✓	$1\frac{1}{4}$ ✓

I can draw a checkmark next to each time after I plot it. That way, I can be sure to plot each time only once.

a. Use the data to complete the line plot below.

Title: _____ *Hours Spent Doing Homework* _____

I can use the title from the table above to write a title for the line plot.

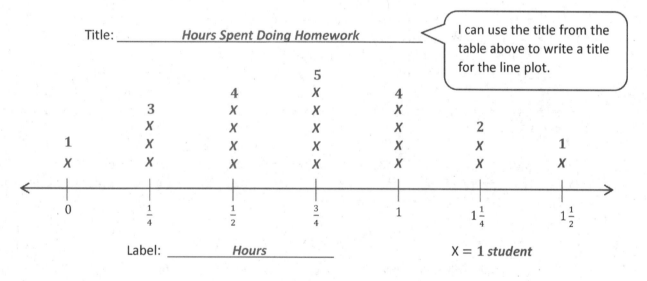

Label: _____ *Hours* _____

X = 1 *student*

b.   How many students spent $\frac{1}{2}$ hour doing their homework?

4 students spent $\frac{1}{2}$ hour doing their homework. < I can count the X's for $\frac{1}{2}$ hour to answer this question.

c.   How many students spent less than 1 hour doing their homework?

13 students spent less than 1 hour doing their homework.

I can count the X's for 0 hours, $\frac{1}{4}$ hour, $\frac{1}{2}$ hour, and $\frac{3}{4}$ hours because these times are all less than 1 hour.

d.   How many students in Mrs. Bishop's class spent time doing homework on Monday night? How do you know?

19 students in Mrs. Bishop's class spent time doing homework on Monday night. I know because I counted all of the X's except the X for 0 hours because that student didn't spend any time doing homework Monday night.

This problem was a little tricky because usually for a problem like this I can just count all of the X's. I can't count all of the X's this time because 1 student spent 0 hours doing homework on Monday night.

e.   Kathleen says most students spent at least 1 hour doing their homework. Is she correct? Explain your thinking.

No, Kathleen is not correct. 7 students spent at least 1 hour doing their homework, but 13 students spent less than 1 hour doing their homework. Kathleen could say that most students spent less than 1 hour doing their homework.

I can count the X's for 1 hour, $1\frac{1}{4}$ hours, and $1\frac{1}{2}$ hours to figure out how many students spent at least 1 hour doing their homework. I can look at my answer to Problem 1(c) to see how many students spent less than 1 hour doing their homework.

© 2018 Great Minds®. eureka-math.org

Name _____ Date _____

Mrs. Felter's students build a model of their school's neighborhood out of blocks. The students measure the heights of the buildings to the nearest $\frac{1}{4}$ inch and record the measurements as shown below.

Heights of Buildings (in Inches)				
$3\frac{1}{4}$	$3\frac{3}{4}$	$4\frac{1}{4}$	$4\frac{1}{2}$	$3\frac{1}{2}$
4	3	$3\frac{3}{4}$	3	$4\frac{1}{2}$
3	$3\frac{1}{2}$	$3\frac{3}{4}$	$3\frac{1}{2}$	4
$3\frac{1}{2}$	$3\frac{1}{4}$	$3\frac{1}{2}$	4	$3\frac{3}{4}$
3	$4\frac{1}{4}$	4	$3\frac{1}{4}$	4

a.   Use the data to complete the line plot below.

Title: _____

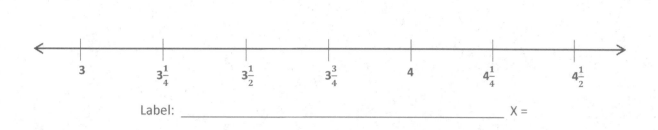

Label: _____   X =

b. How many buildings are $4\frac{1}{4}$ inches tall?

c. How many buildings are less than $3\frac{1}{2}$ inches?

d. How many buildings are in the class model? How do you know?

e. Brook says most buildings in the model are at least 4 inches tall. Is she correct? Explain your thinking.

Samuel is training his frog to compete in the frog-jumping contest at the county fair. The table below shows the distances that Samuel's frog jumped during his training time.

Distance Jumped (in Inches)				
$73\frac{3}{4}$ ✓	74 ✓	$74\frac{1}{4}$ ✓	74 ✓	$73\frac{1}{2}$ ✓
$74\frac{1}{2}$ ✓	$74\frac{1}{4}$ ✓	$74\frac{1}{2}$ ✓	$73\frac{3}{4}$ ✓	74 ✓
$73\frac{1}{4}$ ✓	$\left(74\frac{3}{4}\right)$ ✓	$\left(73\right)$ ✓	$74\frac{1}{4}$ ✓	$73\frac{1}{2}$ ✓
74 ✓	$73\frac{3}{4}$ ✓	74 ✓	74 ✓	$74\frac{1}{4}$ ✓

I can circle the shortest and longest distances to find the endpoints for my line plot.

a. Use the data to create a line plot below.

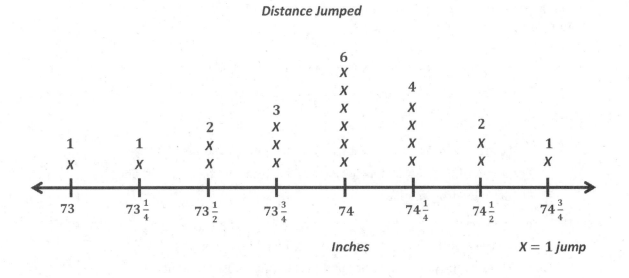

Distance Jumped

Inches   X = 1 jump

b. Explain the steps you took to create the line plot.

I found the endpoints by finding the shortest and longest distances, 73 inches and $74\frac{3}{4}$ inches.

Then I figured out what interval I should use on my line plot by finding the smallest unit, $\frac{1}{4}$ inch.
I marked the endpoints and partitioned and labeled quarter-inch intervals. Then I recorded the data by drawing X's above each measurement. I wrote a title, made a key, and labeled the measurements as Inches.

> I can count by quarter inches from 73 inches to $74\frac{3}{4}$ inches to figure out how many quarter-inch intervals I need on my line plot.

c. How many more times did Samuel's frog jump $74\frac{1}{4}$ inches than $73\frac{1}{2}$ inches?

$4 - 2 = 2$

> I can subtract the number of times the frog jumped $73\frac{1}{2}$ inches from the number of times the frog jumped $74\frac{1}{4}$ inches.

Samuel's frog jumped $74\frac{1}{4}$ inches 2 more times than it jumped $73\frac{1}{2}$ inches.

d. Find the three most frequent measurements on the line plot. What does this tell you about the distance of most of the frog's jumps?

The three most frequent measurements on the line plot are $73\frac{3}{4}$ inches, 74 inches, and $74\frac{1}{4}$
inches. This tells me that most of the frog's jumps were between $73\frac{3}{4}$ inches and $74\frac{1}{4}$ inches.

> I can prove this is true by subtracting the number of times the frog jumped $73\frac{3}{4}$ inches, 74 inches, or $74\frac{1}{4}$ inches from the total number of times the frog jumped.
>
> $20 - 13 = 7$
>
> Thirteen of the frog's jumps were between $73\frac{3}{4}$ inches and $74\frac{1}{4}$ inches. Seven of the jumps were not part of the three most frequent measurements.

Name _____   Date _____

Mrs. Leah's class uses what they learned about simple machines to build marshmallow launchers. They record the distances their marshmallows travel in the chart below.

Distance Traveled (in Inches)				
$48\frac{3}{4}$	49	$49\frac{1}{4}$	50	$49\frac{3}{4}$
$49\frac{1}{2}$	$48\frac{1}{4}$	$49\frac{1}{2}$	$48\frac{3}{4}$	49
$49\frac{1}{4}$	$49\frac{3}{4}$	48	$49\frac{1}{4}$	$48\frac{1}{4}$
49	$48\frac{3}{4}$	49	49	$48\frac{3}{4}$

a.   Use the data to create a line plot below.

b. Explain the steps you took to create the line plot.

c. How many more marshmallows traveled $48\frac{3}{4}$ inches than $48\frac{1}{4}$ inches?

d. Find the three most frequent measurements on the line plot. What does this tell you about the distance that most of the marshmallows traveled?

Lesson 8: Represent measurement data with line plots.

1. The table below shows the amount of money Mrs. Mack's children have in their piggy banks.

Child	Amount of Money
Marie	$16
Nathan	$12
Mara	$15
Noah	$11

Create a picture graph below using the data in the table.

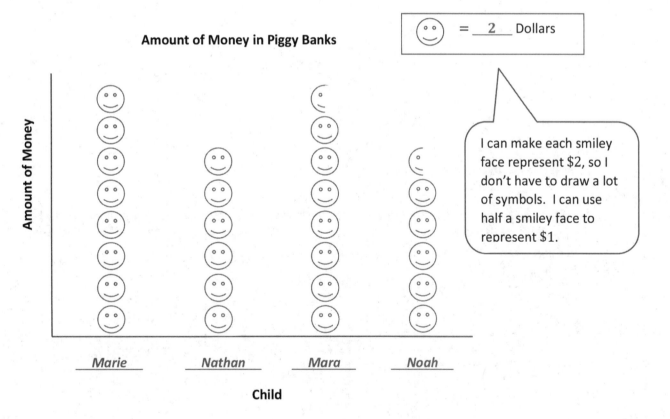

Amount of Money in Piggy Banks

☺ = __2__ Dollars

I can make each smiley face represent $2, so I don't have to draw a lot of symbols. I can use half a smiley face to represent $1.

Amount of Money

Marie   Nathan   Mara   Noah

Child

2. Use the table or graph to answer the following questions.

   a. How much more money do Marie and Nathan have together than Mara and Noah have together?

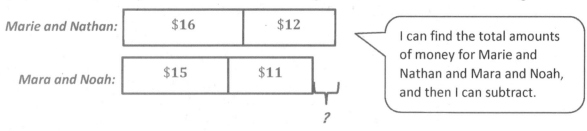

| Marie and Nathan: | $16 | $12 |

I can find the total amounts of money for Marie and Nathan and Mara and Noah, and then I can subtract.

| Mara and Noah: | $15 | $11 |

?

$28 - $26 = $2   *Marie and Nathan have $2 more than Mara and Noah.*

   b. Marie and Noah combine their money to buy packs of baseball cards. Each pack of baseball cards costs $3. How many packs of baseball cards can they buy?

   $16 + $11 = $27

   $27

   | $3 | ? | $3 |
   ...

   I can add to find the total amount of money Marie and Noah have. Then I can divide that amount by $3 to figure out how many packs of baseball cards they can buy.

   $27 ÷ $3 = 9

   Marie and Noah can buy 9 packs of baseball cards.

   c. Mara gets $20 for her birthday. She combines her birthday money with the money in her piggy bank to buy a book for $9 and a bouquet of flowers for her mom. She puts the $8 that she has left back in her piggy bank. How much does the bouquet of flowers cost?

   $20 + $15 = $35

   $35

   | $9 | $8 | ? |

   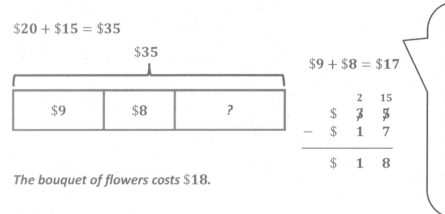

   $9 + $8 = $17

   $$
   \begin{array}{r}
   & \overset{2}{\cancel{3}} \ \overset{15}{\cancel{5}} \\
   \$ & \\
   - \ \$ & 1 \ \ 7 \\
   \hline
   \$ & 1 \ \ 8 \\
   \end{array}
   $$

   I can add the amount of money Mara spends on a book and the amount of money she puts back in her piggy bank. Then, I can subtract that amount from her total amount of money.

   The bouquet of flowers costs $18.

EUREKA MATH®

Name _____ Date _____

1. The table below shows the amount of money Danielle saves for four months.

Month	Money Saved
January	$9
February	$18
March	$36
April	$27

Create a picture graph below using the data in the table.

Money Danielle Saves

= _____ Dollars

Money Saved

Month

2. Use the table or graph to answer the following questions.

    a. How much money does Danielle save in four months?

    b. How much more money does Danielle save in March and April than in January and February?

    c. Danielle combines her savings from March and April to buy books for her friends. Each book costs $9. How many books can she buy?

    d. Danielle earns $33 in June. She buys a necklace for $8 and a birthday present for her brother. She saves the $13 she has left. How much does the birthday present cost?

Lesson 9:   Analyze data to problem solve.

Grade 3
Module 7

1. A museum uses 6 trucks to move paintings and sculptures to a new location. They move a total of 24 paintings and 18 sculptures. Each truck carries an equal number of paintings and an equal number of sculptures. How many paintings and how many sculptures are in each truck?

> I can use the Read-Draw-Write (RDW) process to solve. As I read the problem, I can visualize a picture of the problem in my mind. I know it's helpful to reread the problem in case I missed anything or didn't understand the information completely. Then I can ask myself, "What can I draw?"

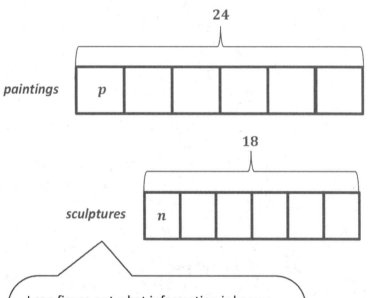

p represents the number of paintings in each truck

$$24 \div 6 = p$$
$$p = 4$$

n represents the number of sculptures in each truck

$$18 \div 6 = n$$
$$n = 3$$

> I can figure out what information is known and unknown using my drawing. I can represent my unknowns using letters. I know there are a total of 24 paintings and 18 sculptures. They are equally placed into 6 trucks. I know the totals and that the number of groups is 6. So my unknown is the size of each group.

> Next, I can write number sentences based on my drawings.

> The final step of the Read-Draw-Write (RDW) process is to write a sentence with words to answer the problem. I can reread the question to be sure that my sentence answers it. This also gives me a chance to look back at my calculation to make sure that my answer is reasonable.

There are 4 paintings and 3 sculptures in each truck.

Lesson 1:   Solve word problems in varied contexts using a letter to represent the unknown.

© 2018 Great Minds®. eureka-math.org

171

2. Christopher's father gives the cashier $30 to pay for 7 keychains from the gift shop. The cashier gives him $9 in change. How much does each keychain cost?

> I know there are many ways to draw and solve this problem, but I want to draw a model that is most helpful to me.

$30

| k | | | | | | | $9 |

$t = \$21$

t represents the total cost of 7 keychains

$\$30 - \$9 = t$
$t = \$21$

k represents the cost of each keychain

$\$21 \div 7 = k$
$k = \$3$

Each keychain costs $3.

> This time I choose to draw only one tape diagram and label both unknowns with letters. I know I first need to solve for *t*, and then I can solve for *k*. Labeling the unknowns with different letters helps me differentiate the two unknowns easily.

> Now I can write my number sentences and a statement that answers the question.

Lesson 1:   Solve word problems in varied contexts using a letter to represent the unknown.

EUREKA MATH

Name _____   Date _____

Max's family takes the train to visit the city zoo. Use the RDW process to solve the problems about Max's trip to the zoo. Use a letter to represent the unknown in each problem.

1. The sign below shows information about the train schedule into the city.

Train Fare—One Way
Adult............... ....$8
Child............. ....$6
Leaves every 15 minutes starting at 6:00 a.m.

   a. Max's family buys 2 adult tickets and 3 child tickets. How much does it cost Max's family to take the train into the city?

   b. Max's father pays for the tickets with $10 bills. He receives $6 in change. How many $10 bills does Max's father use to pay for the train tickets?

   c. Max's family wants to take the fourth train of the day. It's 6:38 a.m. now. How many minutes do they have to wait for the fourth train?

EUREKA
MATH

Lesson 1: Solve word problems in varied contexts using a letter to represent the unknown.

© 2018 Great Minds®. eureka-math.org

173

2. At the city zoo, they see 17 young bats and 19 adult bats. The bats are placed equally into 4 areas. How many bats are in each area?

3. Max's father gives the cashier $20 to pay for 6 water bottles. The cashier gives him $8 in change. How much does each water bottle cost?

4. The zoo has 112 types of reptiles and amphibians in their exhibits. There are 72 types of reptiles, and the rest are amphibians. How many more types of reptiles are there than amphibians in the exhibits?

Lesson 1: Solve word problems in varied contexts using a letter to represent the unknown.

Kathy is 167 centimeters tall. The total height of Kathy and her younger sister Jenny is 319 centimeters. How much taller is Kathy than Jenny? Draw at least 2 different ways to represent the problem.

I can use the RDW process to help me solve. First, I need to read (and reread) the problem. This will help me visualize the problem. Then, I can draw a model to represent the problem with the known and unknown information.

Step 1:

167 cm j cm

Kathy	Jenny

319 cm

j represents Jenny's height in centimeters

$$319 \text{ cm} - 167 \text{ cm} = j$$
$$j = 152 \text{ cm}$$

I notice that this is a two-step problem. From my drawing, I know the total height of the two sisters and the height of Kathy. The unknown in my drawing is Jenny's height, which is labeled with the letter j. I can write a subtraction equation to find her height. But this doesn't answer the question.

Step 2:   167 cm

Kathy

Jenny

d cm

152 cm

d represents the difference between the two heights in centimeters

$$167 \text{ cm} - 152 \text{ cm} = d$$
$$d = 15 \text{ cm}$$

The question is, "How much taller is Kathy than Jenny?" That means I need to draw a second diagram and write a subtraction equation to answer the question. I can label the unknown, which this time is the difference of their heights, with a new letter.

Finally, I can check my work when I write my statement.

Kathy is 15 centimeters taller than Jenny.

EUREKA MATH

Lesson 2: Solve word problems in varied contexts using a letter to represent the unknown.

© 2018 Great Minds®. eureka-math.org

175

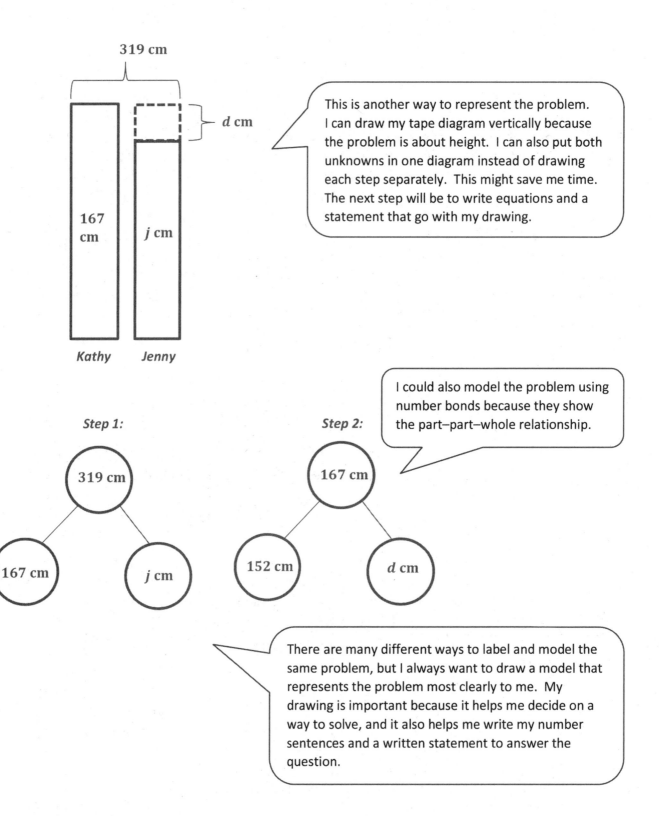

This is another way to represent the problem. I can draw my tape diagram vertically because the problem is about height. I can also put both unknowns in one diagram instead of drawing each step separately. This might save me time. The next step will be to write equations and a statement that go with my drawing.

I could also model the problem using number bonds because they show the part–part–whole relationship.

There are many different ways to label and model the same problem, but I always want to draw a model that represents the problem most clearly to me. My drawing is important because it helps me decide on a way to solve, and it also helps me write my number sentences and a written statement to answer the question.

Lesson 2: Solve word problems in varied contexts using a letter to represent the unknown.

© 2018 Great Minds®. eureka-math.org

Name _____   Date _____

Use the RDW process to solve. Use a letter to represent the unknown in each problem.

1. A box containing 3 small bags of flour weighs 950 grams. Each bag of flour weighs 300 grams. How much does the empty box weigh?

2. Mr. Cullen needs 91 carpet squares. He has 49 carpet squares. If the squares are sold in boxes of 6, how many more boxes of carpet squares does Mr. Cullen need to buy?

3. Erica makes a banner using 4 sheets of paper. Each paper measures 9 inches by 10 inches. What is the total area of Erica's banner?

4. Monica scored 32 points for her team at the Science Bowl. She got 5 four-point questions correct, and the rest of her points came from answering three-point questions. How many three-point questions did she get correct?

5. Kim's black kitten weighs 175 grams. Her gray kitten weighs 43 grams less than the black kitten. What is the total weight of the two kittens?

6. Cassias and Javier's combined height is 267 centimeters. Cassias is 128 centimeters tall. How much taller is Javier than Cassias?

Lesson 2:   Solve word problems in varied contexts using a letter to represent the unknown.

Mrs. Yoon buys 6 bags of counters. Nine counters come in each bag. She gives each of her 12 math students 4 counters. How many counters does she have left?

I will use the RDW (Read-Draw-Write) process to solve this multi-step problem. First I'll read the problem, then I'll pause and visualize what's happening in the problem to get an idea about what to draw.

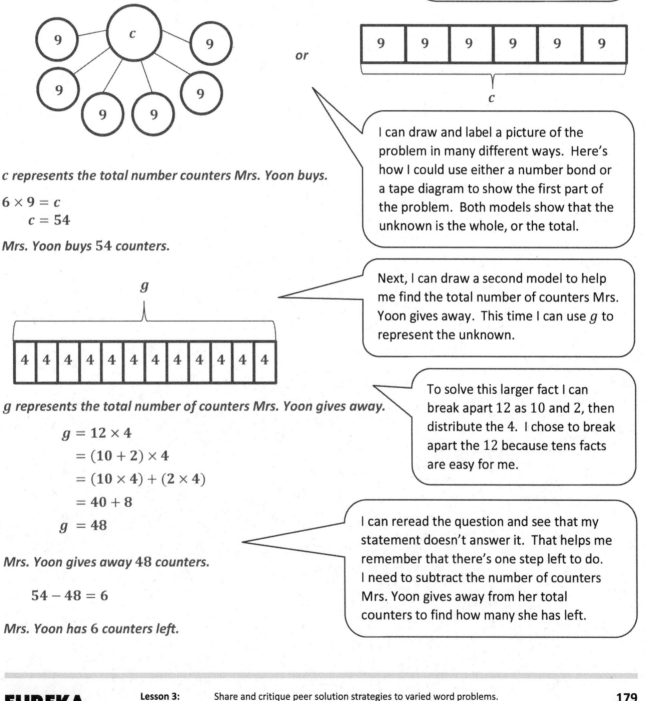

or

c represents the total number counters Mrs. Yoon buys.

$6 \times 9 = c$

$c = 54$

Mrs. Yoon buys 54 counters.

I can draw and label a picture of the problem in many different ways. Here's how I could use either a number bond or a tape diagram to show the first part of the problem. Both models show that the unknown is the whole, or the total.

Next, I can draw a second model to help me find the total number of counters Mrs. Yoon gives away. This time I can use *g* to represent the unknown.

g represents the total number of counters Mrs. Yoon gives away.

$g = 12 \times 4$

$\quad = (10 + 2) \times 4$

$\quad = (10 \times 4) + (2 \times 4)$

$\quad = 40 + 8$

$g = 48$

Mrs. Yoon gives away 48 counters.

$54 - 48 = 6$

Mrs. Yoon has 6 counters left.

To solve this larger fact I can break apart 12 as 10 and 2, then distribute the 4. I chose to break apart the 12 because tens facts are easy for me.

I can reread the question and see that my statement doesn't answer it. That helps me remember that there's one step left to do. I need to subtract the number of counters Mrs. Yoon gives away from her total counters to find how many she has left.

EUREKA
MATH

Lesson 3:   Share and critique peer solution strategies to varied word problems.

179

Name _____   Date _____

Use the RDW process to solve the problems below. Use a letter to represent the unknown in each problem.

1. Jerry pours 86 milliliters of water into 8 tiny beakers. He measures an equal amount of water into the first 7 beakers. He pours the remaining water into the eighth beaker. It measures 16 milliliters. How many milliliters of water are in each of the first 7 beakers?

2. Mr. Chavez's third graders go to gym class at 11:15. Students rotate through three activities for 8 minutes each. Lunch begins at 12:00. How many minutes are there between the end of gym activities and the beginning of lunch?

3. A box contains 100 pens. In each box there are 38 black pens and 42 blue pens. The rest are green pens. Mr. Cane buys 6 boxes of pens. How many green pens does he have in total?

4. Greg has $56. Tom has $17 more than Greg. Jason has $8 less than Tom.

    a. How much money does Jason have?

    b. How much money do the 3 boys have in total?

5. Laura cuts 64 inches of ribbon into two parts and gives her mom one part. Laura's part is 28 inches long. Her mom cuts her ribbon into 6 equal pieces. How long is one of her mom's pieces of ribbon?

1. Complete the chart by answering true or false.

Attribute	Polygon	True or False
Example: 3 Sides		True
Quadrilateral		*True*
2 Sets of Parallel Sides		*False*

> This is true. This shape has four sides and four angles. I know polygons with four straight sides and four angles are called quadrilaterals.

> This is false. This shape only has 1 set of parallel sides. I can think of parallel sides like the two side lines of a capital H, or a slanted *H*, since not all parallel sides stand vertical. Even if the two lines go on forever, they will never cross.

2. Use a straightedge to sketch 2 different quadrilaterals with at least 1 set of parallel sides.

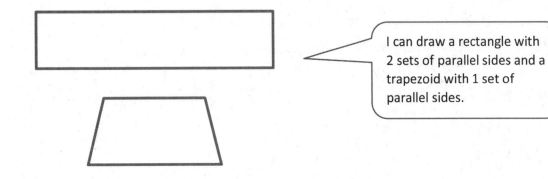

> I can draw a rectangle with 2 sets of parallel sides and a trapezoid with 1 set of parallel sides.

Name _____ Date _____

1. Complete the chart by answering true or false.

Attribute	Polygon	True or False
Example: **3 Sides**		True
4 Sides		
2 Sets of Parallel Sides		
4 Right Angles		
Quadrilateral		

2.   a.   Each quadrilateral below has at least 1 set of parallel sides. Trace each set of parallel sides with a colored pencil.

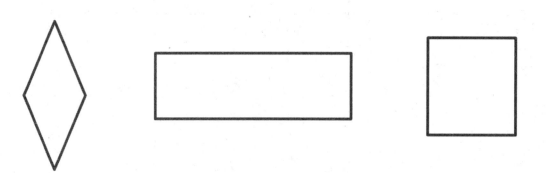

    b.   Using a straightedge, sketch a different quadrilateral with at least 1 set of parallel sides.

1. Match the polygons with their appropriate banners. A polygon can match to more than one banner.

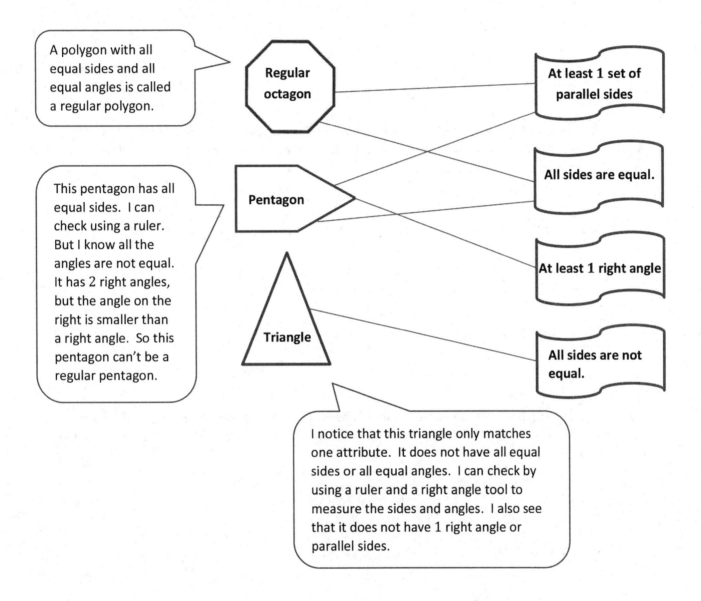

A polygon with all equal sides and all equal angles is called a regular polygon.

This pentagon has all equal sides. I can check using a ruler. But I know all the angles are not equal. It has 2 right angles, but the angle on the right is smaller than a right angle. So this pentagon can't be a regular pentagon.

Regular octagon

Pentagon

Triangle

At least 1 set of parallel sides

All sides are equal.

At least 1 right angle

All sides are not equal.

I notice that this triangle only matches one attribute. It does not have all equal sides or all equal angles. I can check by using a ruler and a right angle tool to measure the sides and angles. I also see that it does not have 1 right angle or parallel sides.

2. Compare the two polygons below. What is the same? What is different?

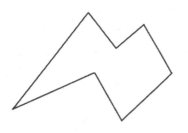 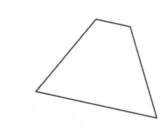

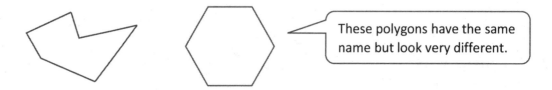

These polygons have the same name but look very different.

Both polygons have 6 sides, so they are both hexagons. The hexagon on the right is a regular hexagon because it has all equal sides and angles. The hexagon on the left does not have all equal sides and angles, so it is not a regular hexagon.

3. David draws the polygons below. Are any of them regular polygons? Explain how you know.

None of David's polygons are regular polygons. I know because I measured the sides and angles of each shape using my ruler and right angle tool, and none of these shapes have all equal sides and all equal angles.

My right angle tool is the corner of an index card. Using my measuring tools helps me to be precise.

Lesson 5:   Compare and classify other polygons.

Name _____   Date _____

1. Match the polygons with their appropriate clouds. A polygon can match to more than 1 cloud.

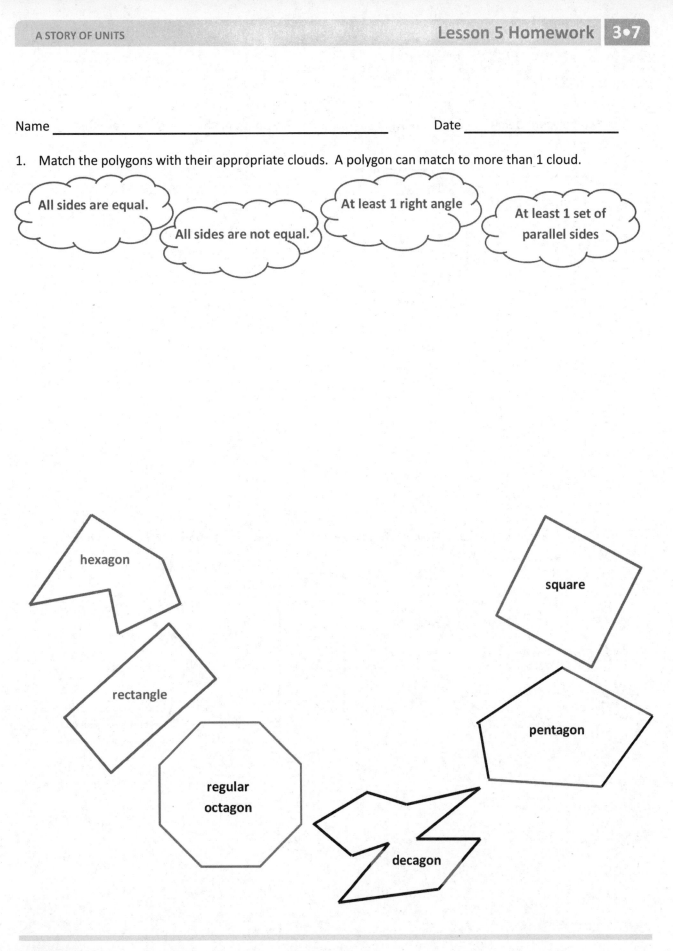

2. The two polygons below are regular polygons. How are these polygons the same? How are they different?

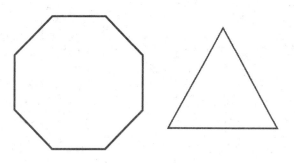

3. Lucia drew the polygons below. Are any of the polygons she drew regular polygons? Explain how you know.

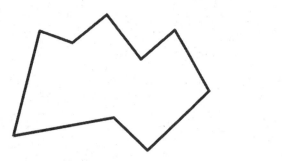  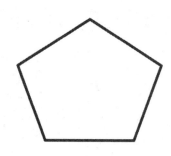

A right angle tool is just the corner of an index card.

Use a ruler and a right angle tool to help you draw the figures with the attributes given below.

1. Draw a triangle with all equal sides.

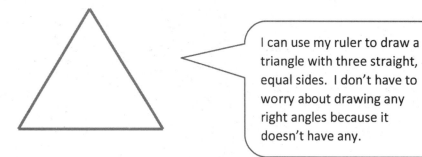

I can use my ruler to draw a triangle with three straight, equal sides. I don't have to worry about drawing any right angles because it doesn't have any.

2. Draw a quadrilateral with at least 1 set of parallel sides and at least 1 right angle. Mark the right angle and parallel sides.

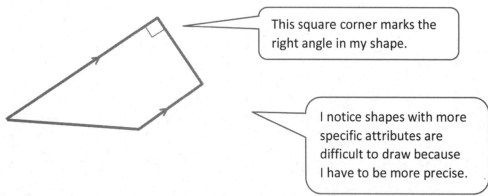

This square corner marks the right angle in my shape.

I notice shapes with more specific attributes are difficult to draw because I have to be more precise.

3. Melissa says she drew a polygon with 4 sides and 4 right angles with no parallel sides. Can Melissa be correct?

Melissa can't be correct because there is no quadrilateral with 4 right angles and no parallel sides. Only rectangles and squares have 4 sides and 4 right angles, but they both have 2 sets of parallel sides.

Name _____   Date _____

Use a ruler and a right angle tool to help you draw the figures with the given attributes below.

1. Draw a triangle that has no right angles.

2. Draw a quadrilateral that has at least 2 right angles.

3. Draw a quadrilateral with 2 equal sides. Label the 2 equal side lengths of your shape.

4. Draw a hexagon with at least 2 equal sides. Label the 2 equal side lengths of your shape.

5. Draw a pentagon with at least 2 equal sides. Label the 2 equal side lengths of your shape.

6. Cristina describes her shape. She says it has 3 equal sides that are each 4 centimeters in length. It has no right angles. Do your best to draw Cristina's shape, and label the side lengths.

Lesson 6: Draw polygons with specified attributes to solve problems.

> The directions tell me the area of each square has to be 16 square units. I can figure out how many tetrominoes I will need by dividing, 16 square units ÷ 4 square units = 4. I will need to use 4 tetrominoes for each square.

1. Use tetrominoes to create three squares, each with an area of 16 square units. Then, color the grid below to show how you created your squares. You may use the same tetromino more than once.

Tetrominoes

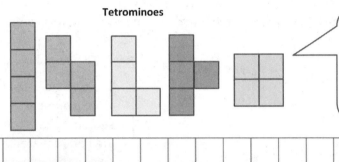

> A tetromino is a shape that has an area of 4 square units, and each square unit shares a whole side with another square unit. This is a set of tetrominoes.

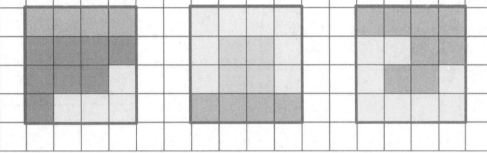

> A strategy I can use to help me make a square with an area of 16 square units is by first marking a 4 by 4 square on the grid. This will help me make sure that my square has the right area. Then I can build the square with the tetrominoes. Sometimes I will need to rotate or flip my tetrominoes to build my shape.

> I can check that my shapes are squares by counting the number of square units on each side and making sure they are all equal. I can also use my right angle tool to make sure that each shape has 4 right angles.

2. Explain how you know the area of each square is 16 square units.

   I know the area of each square is 16 square units because I used 4 tetrominoes to make each square.
   Each tetromino has an area of 4 square units, and 4 × 4 square units = 16 square units.

   a. Write a number sentence to show the area of a square from Problem 1 as the sum of the areas of the tetrominoes you used to make the square.

      Area: 4 square units + 4 square units + 4 square units + 4 square units = 16 square units

   b. Write a number sentence to show the area of a square above as the product of its side lengths.

      Area: 4 units × 4 units = 16 square units

   > The directions say to write a number sentence that shows the area of a square as the sum of the areas of the tetrominoes, so I know that each of my addends is labeled in square units.

   > I know side lengths are measured in length units, and area is labeled in square units.

Lesson 7:   Reason about composing and decomposing polygons using tetrominoes.

EUREKA
MATH

Name _____ Date _____

1. Color tetrominoes on the grid to create three different rectangles. You may use the same tetromino more than once.

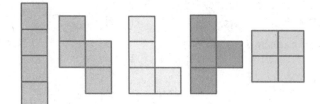

Tetrominoes

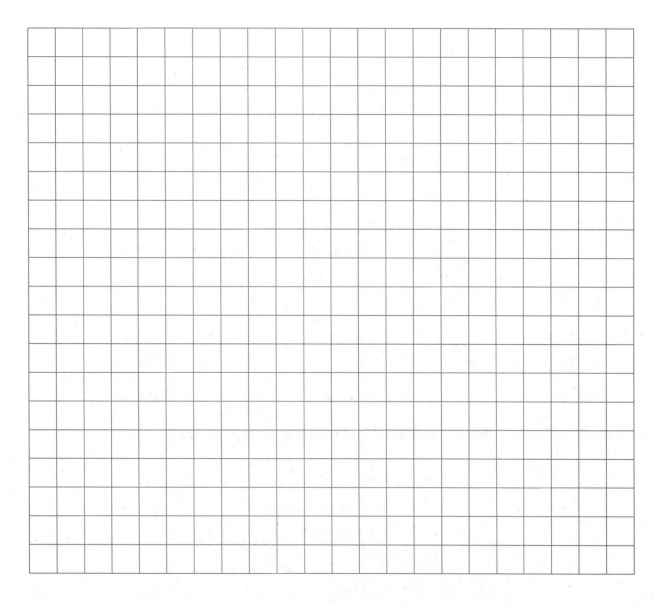

2. Color tetrominoes on the grid below to:

   a. Create a square with an area of 16 square units.

   b. Create at least two different rectangles, each with an area of 24 square units.

   You may use the same tetromino more than once.

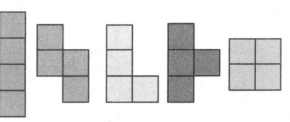

Tetrominoes

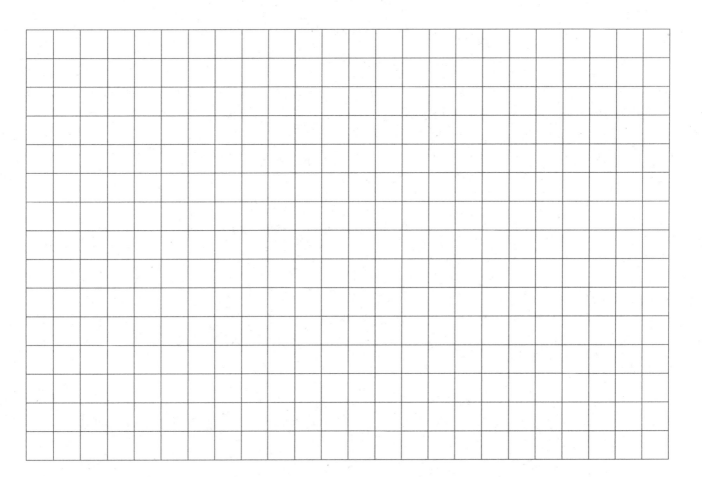

3. Explain how you know the rectangles you created in Problem 2(b) have the correct area.

Lesson 7: Reason about composing and decomposing polygons using tetrominoes.

© 2018 Great Minds®. eureka-math.org

1. Draw a line to divide the rectangle below into 2 equal triangles.

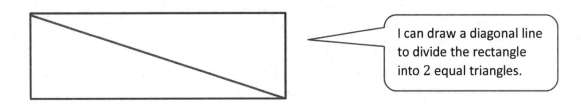

I can draw a diagonal line to divide the rectangle into 2 equal triangles.

2. Draw 2 lines to divide the quadrilateral below into 4 equal triangles.

I can draw 2 diagonal lines to divide this quadrilateral into 4 equal triangles.

3. Choose three shapes from your tangram puzzle. Trace them below. Describe *at least* one attribute that they have in common.

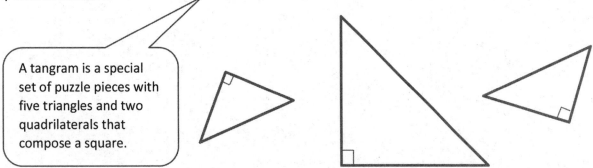

A tangram is a special set of puzzle pieces with five triangles and two quadrilaterals that compose a square.

All three shapes are triangles. They all have 1 right angle and 3 sides. None of the triangles have parallel sides.

Name _____ Date _____

1. Draw a line to divide the square below into 2 equal triangles.

2. Draw a line to divide the triangle below into 2 equal, smaller triangles.

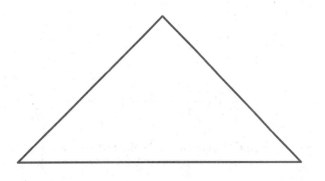

3. Draw a line to divide the trapezoid below into 2 equal trapezoids.

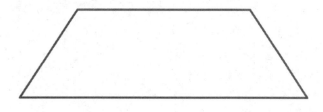

Lesson 8: Create a tangram puzzle and observe relationships among the shapes.

201

4. Draw 2 lines to divide the quadrilateral below into 4 equal triangles.

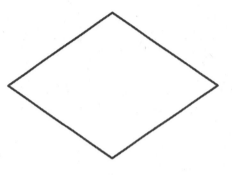

5. Draw 4 lines to divide the square below into 8 equal triangles.

6. Describe the steps you took to divide the square in Problem 5 into 8 equal triangles.

Lesson 8:   Create a tangram puzzle and observe relationships among the shapes.

1. Use your two smallest triangles to create a triangle, a parallelogram, and a square. Show how you created them below.

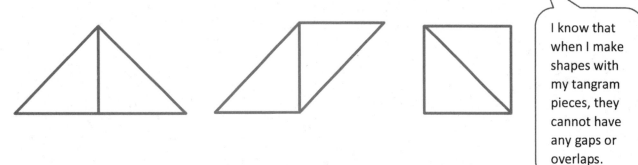

> I know that when I make shapes with my tangram pieces, they cannot have any gaps or overlaps.

2. Use at least two tangram pieces to make and draw as many 4-sided polygons as you can. Draw lines to show where the tangram pieces meet.

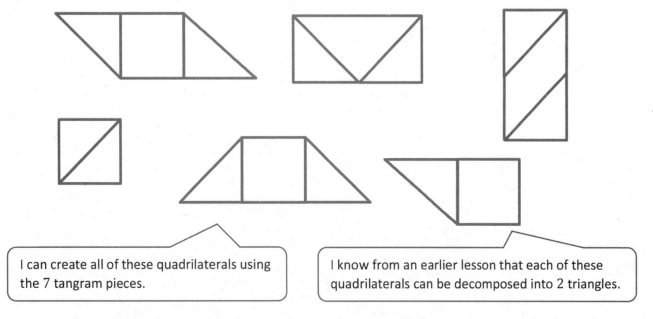

> I can create all of these quadrilaterals using the 7 tangram pieces.

> I know from an earlier lesson that each of these quadrilaterals can be decomposed into 2 triangles.

3. What attributes do your shapes in Problem 2 have in common? What attributes are different?

    All of the shapes I made in Problem 2 are quadrilaterals because they have 4 sides. They all have at least 1 set of parallel lines and 4 angles. Not all of my shapes have equal sides or right angles. That's what makes them different.

Name _____   Date _____

1. Use at least two tangram pieces to make and draw each of the following shapes. Draw lines to show where the tangram pieces meet.

   a. A triangle.

   b. A square.

   c. A parallelogram.

   d. A trapezoid.

2. Use your tangram pieces to create the cat below. Draw lines to show where the tangram pieces meet.

3. Use the five smallest tangram pieces to make a square. Sketch your square below, and draw lines to show where the tangram pieces meet.

1. Trace the perimeter of the shapes below with a black crayon. Then shade in the areas with a blue crayon.

2. Explain how you know you traced the perimeters of the shapes above. How is the perimeter different from the area of a shape?

    I know I traced the perimeters of the shapes because I traced the boundary of each shape with a black crayon, and the boundary is the perimeter. The area of a shape is different than the perimeter. Area measures the amount of space the shape takes up. I shaded the areas of the shapes in blue.

3. Explain how you could use a string to figure out which shape above has the greatest perimeter.

    I can wrap string around each shape and mark where it touches the end after going all around the boundary of the shape. Then I can compare all of the marks, and the shape with the mark farthest from the end of the string has the greatest perimeter.

Lesson 10:   Decompose quadrilaterals to understand perimeter as the boundary of a shape.

207

© 2018 Great Minds®. eureka-math.org

Name _____   Date _____

1. Trace the perimeter of the shapes below.

a. Explain how you know you traced the perimeters of the shapes above.

b. Explain how you could use a string to figure out which shape above has the greatest perimeter.

2. Draw a rectangle on the grid below.

a. Trace the perimeter of the rectangle.

b. Shade the area of the rectangle.

c. How is the perimeter of the rectangle different from the area of the rectangle?

3. Maya draws the shape shown below. Noah colors the inside of Maya's shape as shown. Noah says he colored the perimeter of Maya's shape. Maya says Noah colored the area of her shape. Who is right? Explain your answer.

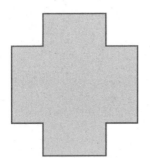

Lesson 10:   Decompose quadrilaterals to understand perimeter as the boundary
                    of a shape.

1. Brian tessellates a parallelogram to make the shape below.

A tessellation is a figure made by copying a shape many times without any gaps or over laps.

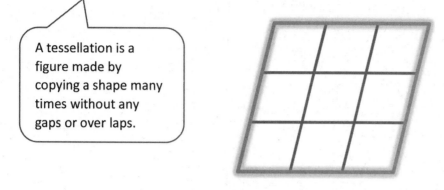

a. Outline the perimeter of Brian's new shape with a highlighter.

b. Name some attributes of his new shape.

    Brian's new shape is a quadrilateral because it has 4 sides. It has 2 sets of parallel lines and 4 angles, but none of them are right angles. Brian created a large parallelogram from smaller parallelograms.

c. Explain how Brian could use a string to measure the perimeter of his new shape.

    Brian could wrap his string around the boundary of his shape and mark where the string touches its end. Then he could measure up to the mark on his string using a ruler.

d. How could Brian increase the perimeter of his tessellation?

    Brian could increase the perimeter of his tessellation by tessellating more shapes. If he tessellated another row or column of shapes, that would increase the perimeter.

I notice that the perimeter of the figure increases with each tessellation and decreases with taking away or erasing tessellations. I know that tessellations could go on forever, even past my paper!

Lesson 11:   Tessellate to understand perimeter as the boundary of a shape.
               (Optional.)   211

© 2018 Great Minds®. eureka-math.org

2. Estimate to draw at least four copies of the given pentagon to make a new shape without gaps or overlaps. Outline the perimeter of your new shape with a highlighter.

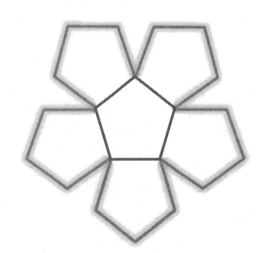

If my tessellations have overlaps or gaps, the shapes wouldn't fit together and the perimeter would not be accurate.

3. The marks on the strings below show the perimeters of Nancy's and Allen's shapes. Whose shape has a greater perimeter? How do you know?

Nancy's String:

Allen's String:

Nancy's shape has a greater perimeter. The mark on the string represents the perimeter of her shape, and it's farther down the string than Allen's mark.

It's just like how I compare numbers on the number line. I can pretend that the end of the string is like zero on the number line. Allen's mark is to the left of Nancy's, so Allen's is smaller because it is a shorter distance from 0.

Lesson 11:   Tessellate to understand perimeter as the boundary of a shape.
                                   (Optional.)

                 © 2018 Great Minds®. eureka-math.org

Name _____   Date _____

1. Samson tessellates regular hexagons to make the shape below.

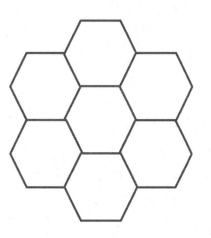

a. Outline the perimeter of Samson's new shape with a highlighter.

b. Explain how Samson could use a string to measure the perimeter of his new shape.

c. How many sides does his new shape have?

d. Shade in the area of his new shape with a colored pencil.

2. Estimate to draw at least four copies of the given triangle to make a new shape, without gaps or overlaps. Outline the perimeter of your new shape with a highlighter. Shade in the area with a colored pencil.

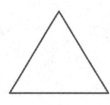

Lesson 11:   Tessellate to understand perimeter as the boundary of a shape.   213
              (Optional.)

© 2018 Great Minds®. eureka-math.org

3. The marks on the strings below show the perimeters of Shyla's and Frank's shapes. Whose shape has a greater perimeter? How do you know?

Shyla's String:

Frank's String:

4. India and Theo use the same shape to create the tessellations shown below.

India's Tessellation

Theo's Tessellation

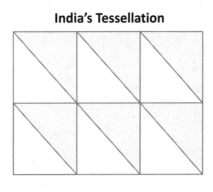

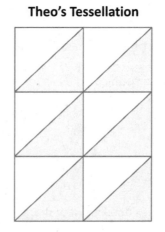

a. Estimate to draw the shape India and Theo used to make their tessellations.

b. Theo says both tessellations have the same perimeter. Do you think Theo is right? Why or why not?

214   Lesson 11:   Tessellate to understand perimeter as the boundary of a shape.
(Optional.)

© 2018 Great Minds®. eureka-math.org

1. Measure and label the side lengths of the shapes below in centimeters. Then, find the perimeter of each shape.

a.

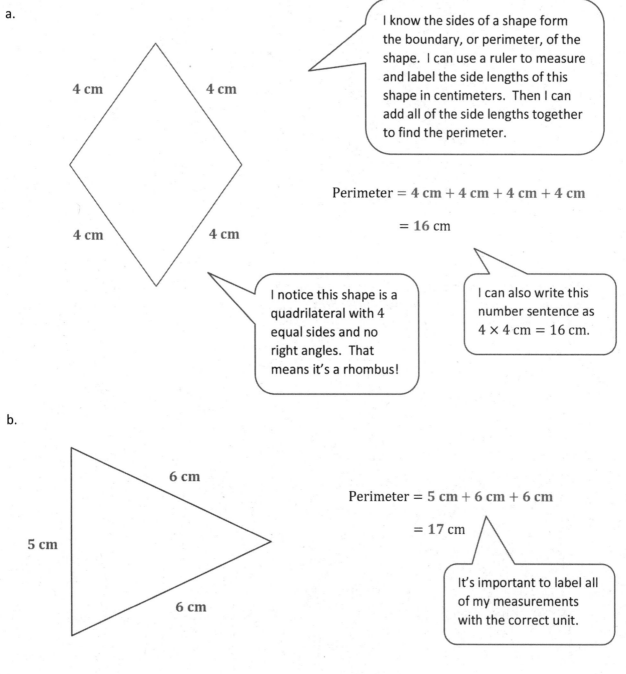

> I know the sides of a shape form the boundary, or perimeter, of the shape. I can use a ruler to measure and label the side lengths of this shape in centimeters. Then I can add all of the side lengths together to find the perimeter.

Perimeter $= 4$ cm $+ 4$ cm $+ 4$ cm $+ 4$ cm

$= 16$ cm

> I notice this shape is a quadrilateral with 4 equal sides and no right angles. That means it's a rhombus!

> I can also write this number sentence as 4×4 cm $= 16$ cm.

b.

Perimeter $= 5$ cm $+ 6$ cm $+ 6$ cm

$= 17$ cm

> It's important to label all of my measurements with the correct unit.

Lesson 12: Measure side lengths in whole number units to determine the perimeter of polygons.

215

2. Albert measures the two side lengths of the rectangle shown below. He says he can find the perimeter with the measurements. Explain Albert's thinking. Then, find the perimeter in centimeters.

8 cm

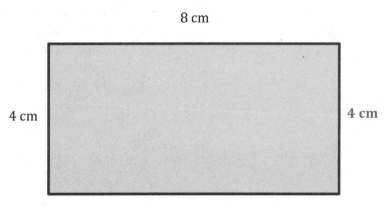

4 cm   4 cm

8 cm

Albeit can find the perimeter using the two side lengths he measured because opposite sides of a rectangle are equal. Since he knows the lengths of the two sides, he knows the lengths of the other two sides. Now he can find the perimeter.

Perimeter = 4 cm + 8 cm + 4 cm + 8 cm

= 24 cm

I can also think of this problem as
3 eights = 24, or 12 + 12 = 24.

The perimeter of the rectangle is 24 centimeters

Lesson 12:   Measure side lengths in whole number units to determine the
perimeter of polygons.

Name _____   Date _____

1. Measure and label the side lengths of the shapes below in centimeters. Then, find the perimeter of each shape.

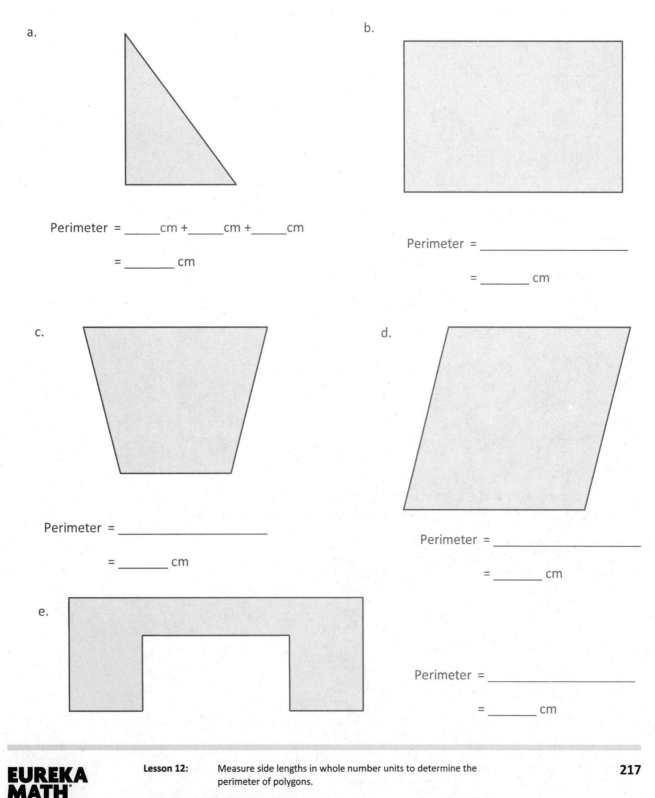

a.

Perimeter = _____cm + _____cm + _____cm

= _____ cm

b.

Perimeter = _____

= _____ cm

c.

Perimeter = _____

= _____ cm

d.

Perimeter = _____

= _____ cm

e.

Perimeter = _____

= _____ cm

2. Melinda draws two trapezoids to create the hexagon shown below. Use a ruler to find the side lengths of Melinda's hexagon in centimeters. Then, find the perimeter.

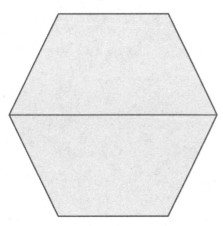

3. Victoria and Eric draw the shapes shown below. Eric says his shape has a greater perimeter because it has more sides than Victoria's shape. Is Eric right? Explain your answer.

Victoria's Shape   **Eric's Shape**

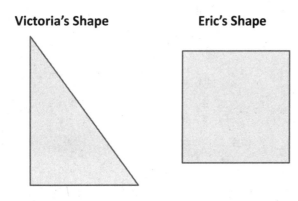

4. Jamal uses his ruler and a right angle tool to draw the rectangle shown below. He says the perimeter of his rectangle is 32 centimeters. Do you agree with Jamal? Why or why not?

Lesson 12:   Measure side lengths in whole number units to determine the perimeter of polygons.

© 2018 Great Minds®. eureka-math.org

1. Find the perimeter of the following shapes.

> I see that the side lengths of each shape are already given, so I do not need to measure them. Now I just need to add the side lengths to find the perimeter.

a.

3 in

5 in   5 in

7 in

$P = 3\text{ in} + 5\text{ in} + 5\text{ in} + 7\text{ in}$

$P = 20\text{ in}$

> This quadrilateral has 1 set of parallel lines and no right angles. It's a trapezoid.

b.

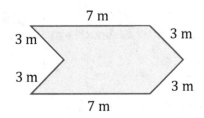

7 m

3 m   3 m

3 m

3 m

7 m

$P = 3\text{ m} + 3\text{ m} + 7\text{ m} + 3\text{ m} + 7\text{ m} + 3\text{ m}$

$P = 26\text{ m}$

> This shape has six sides, so it's a hexagon. It is not a regular hexagon because it does not have all equal sides.

> I notice that each shape uses different units to measure. I need to make sure to label my measurements and their units correctly.

2. Allyson's rectangular garden is 31 feet long and 49 feet wide. What is the perimeter of Allyson's garden?

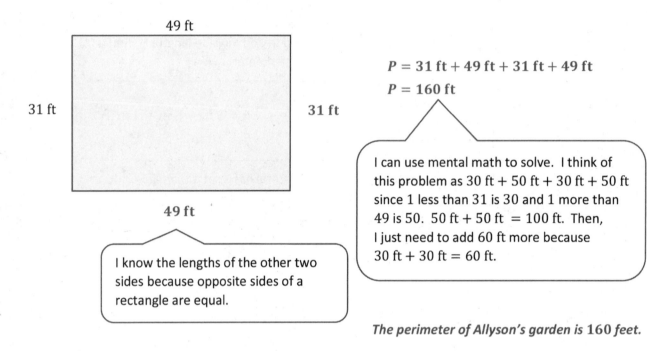

49 ft

31 ft   31 ft

49 ft

$P = 31\ \text{ft} + 49\ \text{ft} + 31\ \text{ft} + 49\ \text{ft}$
$P = 160\ \text{ft}$

I can use mental math to solve. I think of this problem as 30 ft + 50 ft + 30 ft + 50 ft since 1 less than 31 is 30 and 1 more than 49 is 50. 50 ft + 50 ft = 100 ft. Then, I just need to add 60 ft more because 30 ft + 30 ft = 60 ft.

I know the lengths of the other two sides because opposite sides of a rectangle are equal.

The perimeter of Allyson's garden is 160 feet.

EUREKA
MATH®

Name _____ Date _____

1. Find the perimeters of the shapes below. Include the units in your equations. Match the letter inside each shape to its perimeter to solve the riddle. The first one has been done for you.

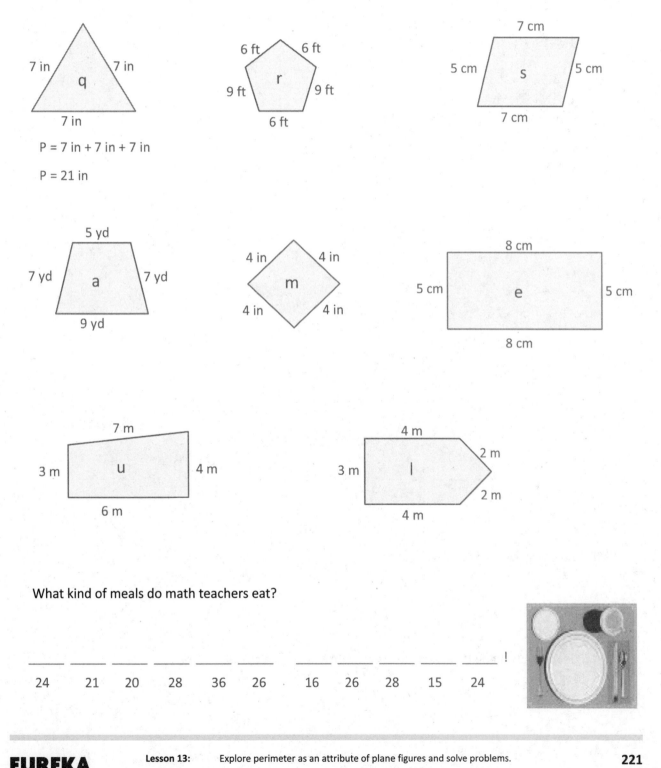

P = 7 in + 7 in + 7 in

P = 21 in

What kind of meals do math teachers eat?

___ ___ ___ ___ ___ ___   ___ ___ ___ ___ ___ !
24   21   20   28   36   26   16   26   28   15   24

2. Alicia's rectangular garden is 33 feet long and 47 feet wide. What is the perimeter of Alicia's garden?

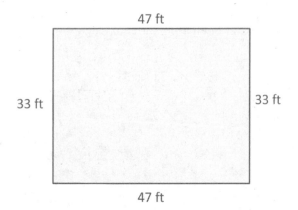

47 ft

33 ft   33 ft

47 ft

3. Jaques measured the side lengths of the shape below.

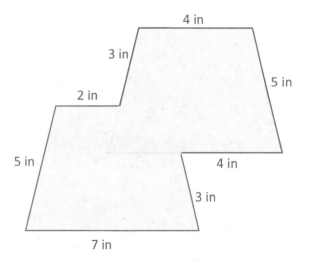

4 in

3 in

2 in

5 in

5 in

4 in

3 in

7 in

a. Find the perimeter of Jaques's shape.

b. Jaques says his shape is an octagon. Is he right? Why or why not?

EUREKA MATH

1. Label the unknown side lengths of the regular shapes below. Then, find the perimeter of each shape.

a.

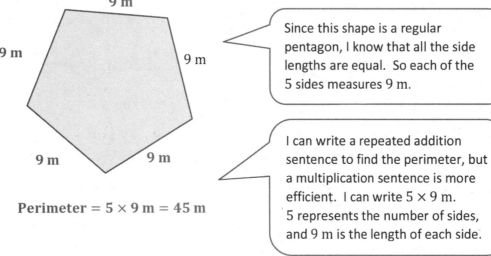

Since this shape is a regular pentagon, I know that all the side lengths are equal. So each of the 5 sides measures 9 m.

Perimeter = 5 × 9 m = 45 m

I can write a repeated addition sentence to find the perimeter, but a multiplication sentence is more efficient. I can write 5 × 9 m. 5 represents the number of sides, and 9 m is the length of each side.

b.

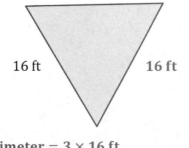

16 ft

16 ft   16 ft

Perimeter = 3 × 16 ft

= (3 × 10 ft) + (3 × 6 ft)

= 30 ft + 18 ft

= 48 ft

I can use the break apart and distribute strategy to solve for a large fact like 3 × 16 ft. I can break apart 16 ft as 10 ft and 6 ft since multiplying by tens is easy. Then I can add the two smaller facts to find the answer to the larger fact.

Lesson 14:   Determine the perimeter of regular polygons and rectangles when   223
              whole number measurements are unknown.

© 2018 Great Minds®. eureka-math.org

2. Jake traces a regular octagon on his paper. Each side measures 6 centimeters. He also traces a regular decagon on his paper. Each side of the decagon measures 4 centimeters. Which shape has a greater perimeter? Show your work.

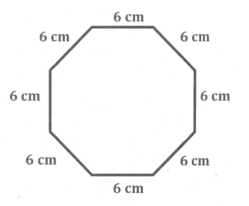

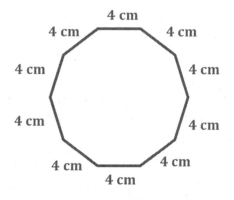

Perimeter = 8 × 6 cm = 48 cm   **Perimeter = 10 × 4 cm = 40 cm**

Jake's octagon has a greater perimeter by 8 cm.

Even though a decagon has more sides than an octagon, the side lengths of Jake's octagon are longer than the side lengths of his decagon. That's why Jake's octagon has a greater perimeter.

Lesson 14:   Determine the perimeter of regular polygons and rectangles when
                                whole number measurements are unknown.

Name _____ Date _____

1. Label the unknown side lengths of the regular shapes below. Then, find the perimeter of each shape.

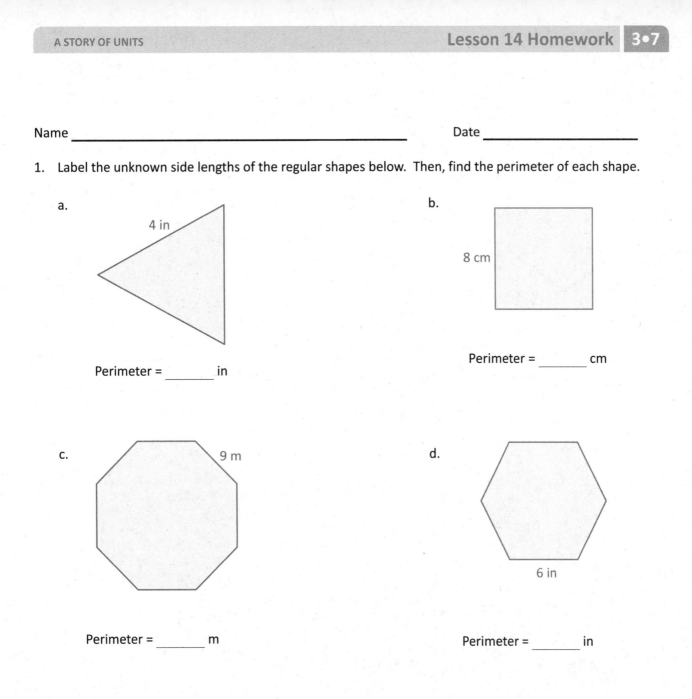

a.

4 in

Perimeter = _____ in

b.

8 cm

Perimeter = _____ cm

c.

9 m

Perimeter = _____ m

d.

6 in

Perimeter = _____ in

2. Label the unknown side lengths of the rectangle below. Then, find the perimeter of the rectangle.

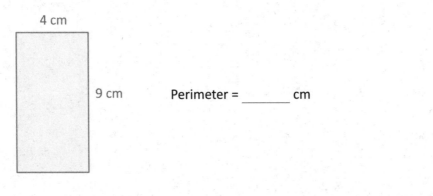

4 cm

9 cm   Perimeter = _____ cm

EUREKA MATH

Lesson 14: Determine the perimeter of regular polygons and rectangles when whole number measurements are unknown.

225

© 2018 Great Minds®. eureka-math.org

3. Roxanne draws a regular pentagon and labels a side length as shown below. Find the perimeter of Roxanne's pentagon.

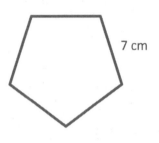

7 cm

4. Each side of a square field measures 24 meters. What is the perimeter of the field?

5. What is the perimeter of a rectangular sheet of paper that measures 8 inches by 11 inches?

Lesson 14:   Determine the perimeter of regular polygons and rectangles when
whole number measurements are unknown.

1. Mr. Kim builds a 7 ft by 9 ft rectangular fence around his vegetable garden. What is the total length of Mr. Kim's fence?

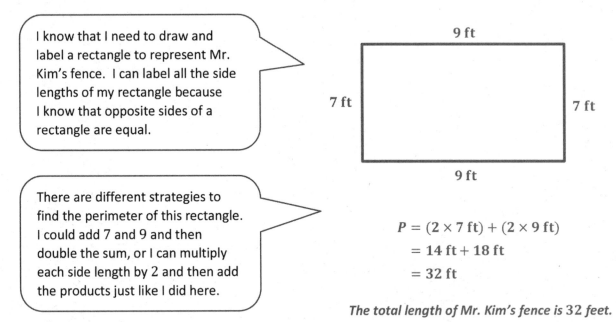

I know that I need to draw and label a rectangle to represent Mr. Kim's fence. I can label all the side lengths of my rectangle because I know that opposite sides of a rectangle are equal.

There are different strategies to find the perimeter of this rectangle. I could add 7 and 9 and then double the sum, or I can multiply each side length by 2 and then add the products just like I did here.

$$P = (2 \times 7 \text{ ft}) + (2 \times 9 \text{ ft})$$
$$= 14 \text{ ft} + 18 \text{ ft}$$
$$= 32 \text{ ft}$$

The total length of Mr. Kim's fence is 32 feet.

2. Gracie uses regular triangles to make the shape below. Each side length of a triangle measures 4 cm. What is the perimeter of Grade's shape?

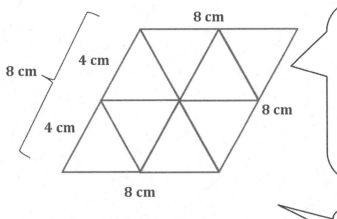

I know that each side length of the regular triangle is 4 cm. Since each side length of Gracie's larger shape is made up of 2 sides of a triangle, the side length of the larger shape is 8 cm. Now I can find the perimeter of her shape by writing a repeated addition sentence or multiplying the 4 side lengths by 8 cm.

$$P = 4 \times 8 \text{ cm} = 32 \text{ cm}$$

The perimeter of Gracie's shape is 32 cm.

Gracie's new shape has 4 equal sides and no right angles. It's a rhombus!

Name _____   Date _____

1. Miguel glues a ribbon border around the edges of a 5-inch by 8-inch picture to create a frame. What is the total length of ribbon Miguel uses?

2. A building at Elmira College has a room shaped like a regular octagon. The length of each side of the room is 5 feet. What is the perimeter of this room?

3. Manny fences in a rectangular area for his dog to play in the backyard. The area measures 35 yards by 45 yards. What is the total length of fence that Manny uses?

4. Tyler uses 6 craft sticks to make a hexagon. Each craft stick is 6 inches long. What is the perimeter of Tyler's hexagon?

5. Francis made a rectangular path from her driveway to the porch. The width of the path is 2 feet. The length is 28 feet longer than the width. What is the perimeter of the path?

6. The gym teacher uses tape to mark a 4-square court on the gym floor as shown. The outer square has side lengths of 16 feet. What is the total length of tape the teacher uses to mark Square A?

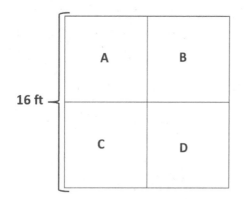

16 ft

1. Alicia draws the shape below.

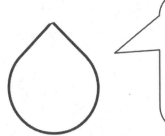

> I know shapes that don't have straight lines, like circles, still have a perimeter. But I can't just use rulers to find their perimeters. I can estimate by using a string to represent the perimeter and then measure the string.

a. Explain how Alicia could use string and a ruler to find the shape's perimeter.

   Alicia can wrap string around the boundary of her shape. Then, she can mark where the string meets the end after going all the way around once. Finally, she can use a ruler to measure from the end of the string to the mark.

   > I know this method does not give me an exact perimeter since I am using string. It is a close estimate.

b. Would you use this method to find the perimeter of a rectangle? Explain why or why not.

   I would not use this method to find the perimeter of a rectangle. Using string is not as efficient or as precise as measuring the sides of a rectangle with a ruler and then adding the side lengths together.

2. Can you find the perimeter of the shape below using just your ruler? Explain your answer.

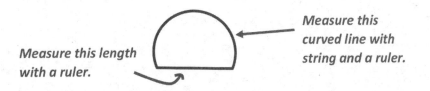

Measure this length with a ruler.

Measure this curved line with string and a ruler.

No, I can't find the perimeter of the shape using just my ruler. The boundary of the shape has a curved line, and I can't measure curved lines with just a ruler. I can measure the straight side length with a ruler and use string to measure the curved line. Then, I can add the two measurements together to find the perimeter.

Lesson 16:   Use string to measure the perimeter of various circles to the nearest quarter inch.

231

© 2018 Great Minds®. eureka-math.org

Name _____ Date _____

1. a. Find the perimeter of 5 circular objects from home to the nearest quarter inch using string. Record the name and perimeter of each object in the chart below.

Object	Perimeter (to the nearest quarter inch)
Example: Peanut Butter Jar Cap	$9\frac{1}{2}$ inches

   b. Explain the steps you used to find the perimeter of the circular objects in the chart above.

Lesson 16: Use string to measure the perimeter of various circles to the nearest quarter inch.

233

2. Use your string and ruler to find the perimeter of the two shapes below to the nearest quarter inch.

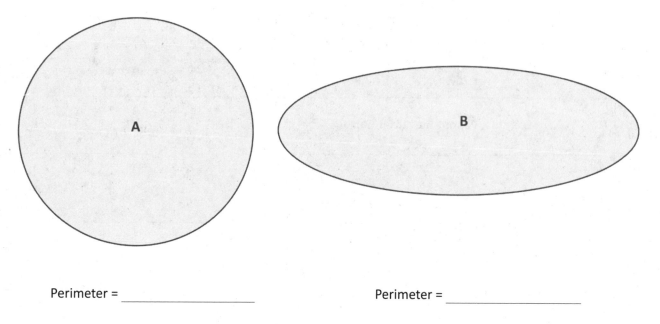

Perimeter = _____   Perimeter = _____

a. Which shape has a greater perimeter?

b. Find the difference between the two perimeters.

3. Describe the steps you took to find the perimeter of the objects in Problem 2. Would you use this method to find the perimeter of a square? Explain why or why not.

Lesson 16:   Use string to measure the perimeter of various circles to the nearest quarter inch.

1. The shape below is made up of rectangles. Label the unknown side lengths. Then, write and solve an equation to find the perimeter of the shape.

This is one way I can visualize how two rectangles fit together to make this shape.

If I extended the line on the bottom to match the one at the top, it would be 6 cm because opposite sides of a rectangle are equal. Knowing that, I can subtract the part labeled 3 cm from 6 cm to find the length of the bottom line.

I can find this unknown side length by adding the known widths, 3 cm and 2 cm, to get 5 cm. This whole side length is 5 cm.

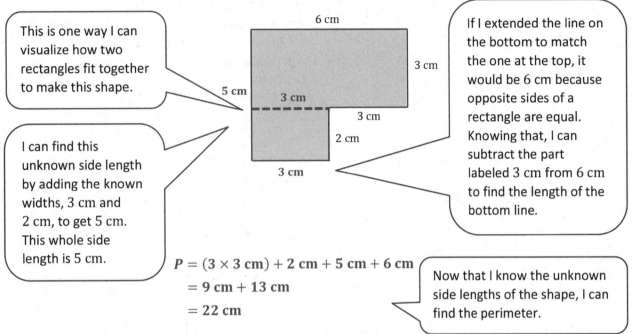

$P = (3 \times 3 \text{ cm}) + 2 \text{ cm} + 5 \text{ cm} + 6 \text{ cm}$
$\quad = 9 \text{ cm} + 13 \text{ cm}$
$\quad = 22 \text{ cm}$

Now that I know the unknown side lengths of the shape, I can find the perimeter.

The perimeter of the shape is 22 cm

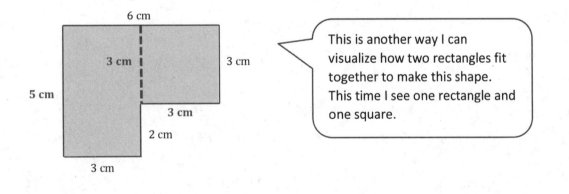

This is another way I can visualize how two rectangles fit together to make this shape. This time I see one rectangle and one square.

Lesson 17:   Use all four operations to solve problems involving perimeter and
unknown measurements.

© 2018 Great Minds®. eureka-math.org

2. Label the unknown side lengths. Then, find the perimeter of the shaded rectangle.

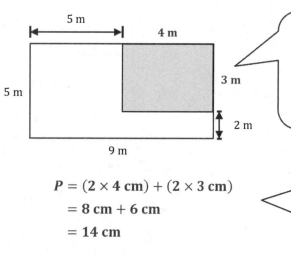

I know the side lengths of the whole rectangle are 9 m and 5 m. In order to find the side lengths of the shaded part, I can subtract the total lengths from the known parts.
9 m − 5 m = 4 m, and 5 m − 2 m = 3 m.

$$P = (2 \times 4 \text{ cm}) + (2 \times 3 \text{ cm})$$
$$= 8 \text{ cm} + 6 \text{ cm}$$
$$= 14 \text{ cm}$$

The perimeter of the shaded rectangle is 14 **cm.**

Now that I know the side lengths of the shaded part, I can find the perimeter.
I know from the question that the shaded part is a rectangle. So it's opposite sides are equal.

Lesson 17:   Use all four operations to solve problems involving perimeter and
                      unknown measurements.

© 2018 Great Minds®. eureka-math.org

EUREKA
MATH®

Name _____ Date _____

1. The shapes below are made up of rectangles. Label the unknown side lengths. Then, write and solve an equation to find the perimeter of each shape.

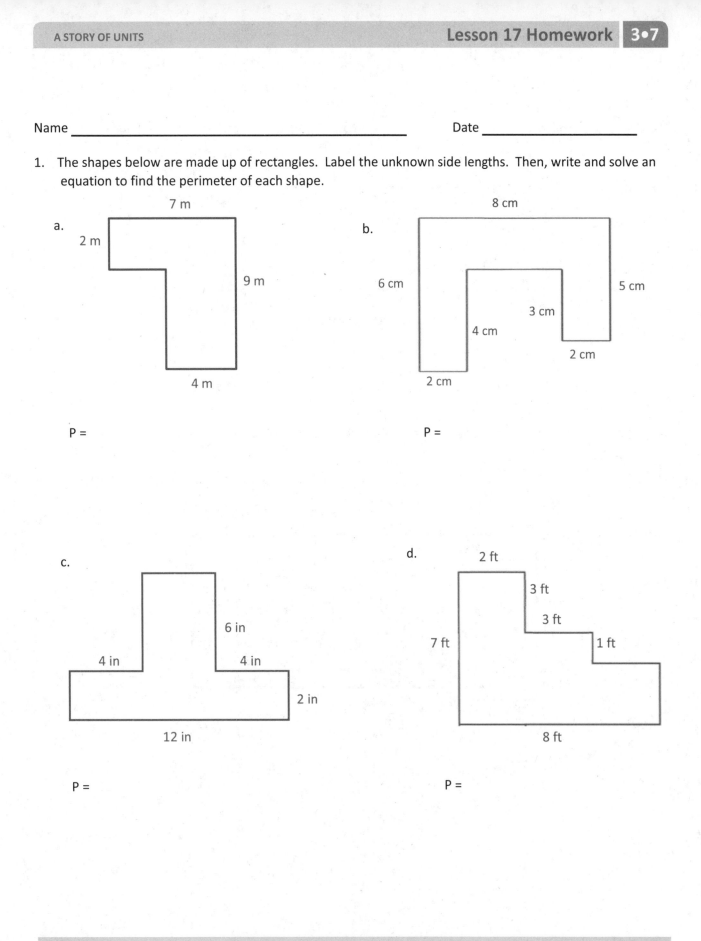

a.

7 m

2 m

9 m

4 m

P =

b.

8 cm

6 cm

4 cm

3 cm

5 cm

2 cm

2 cm

P =

c.

6 in

4 in

4 in

2 in

12 in

P =

d.

2 ft

3 ft

3 ft

7 ft

1 ft

8 ft

P =

Lesson 17: Use all four operations to solve problems involving perimeter and unknown measurements.

237

© 2018 Great Minds®. eureka-math.org

2. Sari draws and labels the squares and rectangle below. Find the perimeter of the new shape.

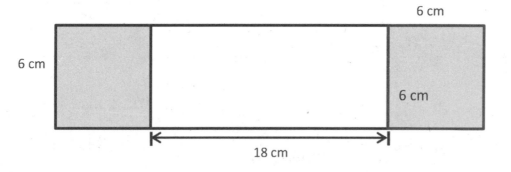

3. Label the unknown side lengths. Then, find the perimeter of the shaded rectangle.

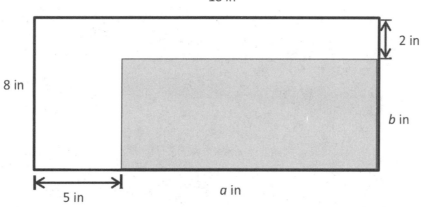

Lesson 17: Use all four operations to solve problems involving perimeter and unknown measurements.

© 2018 Great Minds®. eureka-math.org

Estimate to draw as many rectangles as you can with an area of 15 square centimeters. Label the side lengths of each rectangle.

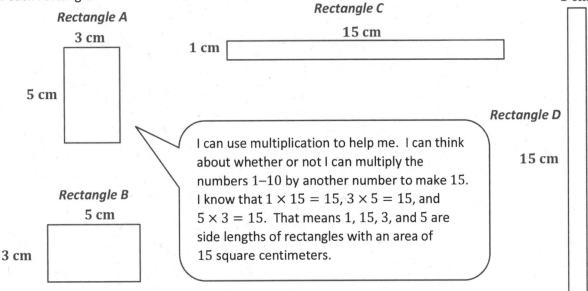

a. Which rectangles above have the greatest perimeter? How do you know just by looking at their shape?

Rectangles C and D have the greatest perimeter. They both have a perimeter of 32 centimeters. I can tell just by looking at their shapes that they have the greatest perimeter because they are longer and skinnier than Rectangles A and B.

I know that long, skinny rectangles have larger perimeters than short, wide rectangles with the same area. Long side lengths add up to greater perimeters than short side lengths.

b. Which rectangles above have the smallest perimeter? How do you know just by looking at their shape?

Rectangles A and B have the smallest perimeter. They both have a perimeter of 16 centimeters. I can tell just by looking at their shapes that they have the smallest perimeter because they are shorter and wider than Rectangles C and D.

I know that short, wide rectangles have smaller perimeters than long, skinny rectangles with the same area. Short side lengths add up to smaller perimeters than long side lengths.

Lesson 18:   Construct rectangles from a given number of unit squares and determine the perimeters.

© 2018 Great Minds®. eureka-math.org

239

Name _____   Date _____

1. Shade in squares on the grid below to create as many rectangles as you can with an area of 18 square centimeters.

2. Find the perimeter of each rectangle in Problem 1 above.

Lesson 18:   Construct rectangles from a given number of unit squares and
               determine the perimeters.

© 2018 Great Minds®. eureka-math.org

241

3. Estimate to draw as many rectangles as you can with an area of 20 square centimeters. Label the side lengths of each rectangle.

a. Which rectangle above has the greatest perimeter? How do you know just by looking at its shape?

b. Which rectangle above has the smallest perimeter? How do you know just by looking at its shape?

Lesson 18:   Construct rectangles from a given number of unit squares and determine the perimeters.

1. Use unit squares to make rectangles for each given number below. Complete the charts to show how many rectangles you can make for each given number of unit squares. You might not use all the spaces in each chart.

Number of unit squares = 12	
Number of rectangles I made: __3__	
Width	Length
1	12
2	6
3	4

Number of unit squares = 13	
Number of rectangles I made: __1__	
Width	Length
1	13

Number of unit squares = 14	
Number of rectangles I made: __2__	
Width	Length
1	14
2	7

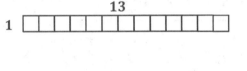

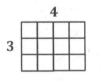

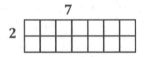

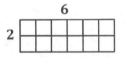

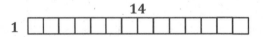

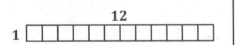

I can use multiplication to help me. I can think about whether or not I can multiply the numbers 1–10 by another number to make 12, 13, or 14. Once I figure out factors that equal those numbers when multiplied, I can build rectangles with the factors as the side lengths.

Lesson 19:   Use a line plot to record the number of rectangles constructed from a given number of unit squares.

243

2. Create a line plot with the data you collected in Problem 1.

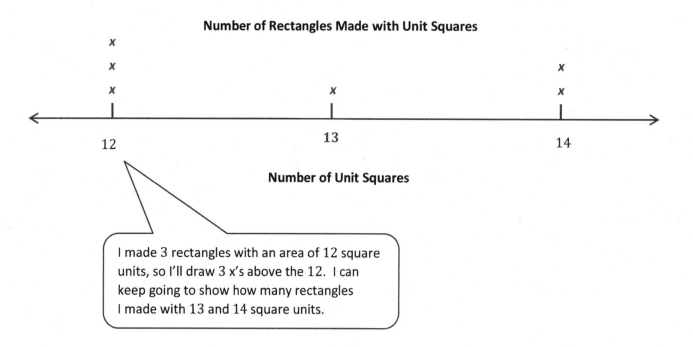

Number of Rectangles Made with Unit Squares

12   13   14

Number of Unit Squares

I made 3 rectangles with an area of 12 square units, so I'll draw 3 x's above the 12. I can keep going to show how many rectangles I made with 13 and 14 square units.

Lesson 19:   Use a line plot to record the number of rectangles constructed from a given number of unit squares.

EUREKA MATH®

Name _____ Date _____

1. Cut out the unit squares at the bottom of the page. Then, use them to make rectangles for each given number of unit squares. Complete the charts to show how many rectangles you can make for each given number of unit squares. You might not use all the spaces in each chart.

Number of unit squares = **6**

Number of rectangles
I made: _____

Width	Length

Number of unit squares = **7**

Number of rectangles
I made: _____

Width	Length

Number of unit squares = **8**

Number of rectangles
I made: _____

Width	Length

Number of unit squares = **9**

Number of rectangles
I made: _____

Width	Length

Number of unit squares = **10**

Number of rectangles
I made: _____

Width	Length

Number of unit squares = **11**

Number of rectangles
I made: _____

Width	Length

✄ -

  Lesson 19:   Use a line plot to record the number of rectangles constructed from a given number of unit squares.

245

© 2018 Great Minds®. eureka-math.org

2. Create a line plot with the data you collected in Problem 1.

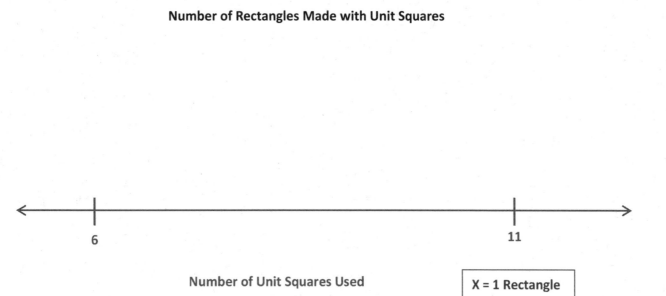

Number of Rectangles Made with Unit Squares

Number of Unit Squares Used

X = 1 Rectangle

a. Luke looks at the line plot and says that all odd numbers of unit squares produce only 1 rectangle. Do you agree? Why or why not?

b. How many X's would you plot for 4 unit squares? Explain how you know.

Lesson 19:   Use a line plot to record the number of rectangles constructed from a given number of unit squares.

© 2018 Great Minds®. eureka-math.org

247

1. Rex uses unit square tiles to make rectangles with a perimeter of 12 units. He draws his rectangles as shown below. Can Rex make another rectangle using unit square tiles that has a perimeter of 12 units? Explain your answer.

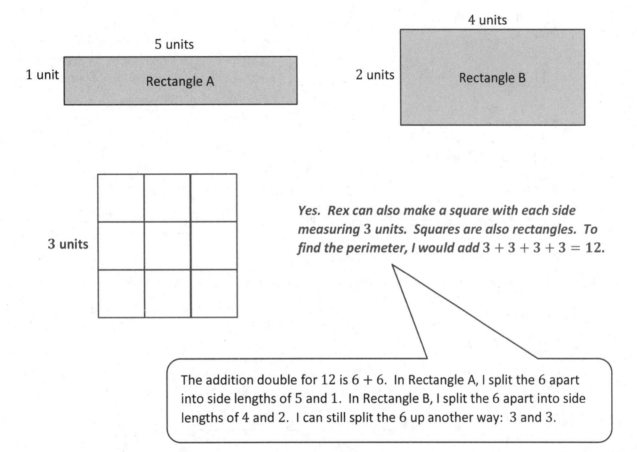

Yes. Rex can also make a square with each side measuring 3 units. Squares are also rectangles. To find the perimeter, I would add $3 + 3 + 3 + 3 = 12$.

The addition double for 12 is $6 + 6$. In Rectangle A, I split the 6 apart into side lengths of 5 and 1. In Rectangle B, I split the 6 apart into side lengths of 4 and 2. I can still split the 6 up another way: 3 and 3.

Lesson 20:   Construct rectangles with a given perimeter using unit squares and determine their areas.

249

© 2018 Great Minds®. eureka-math.org

2. Maureen draws a square that has a perimeter of 24 centimeters.

   a. Estimate to draw Maureen's square below. Label the length and width of the square.

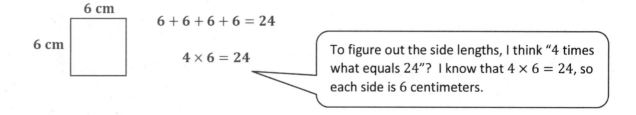

   6 cm

   6 cm

   $6 + 6 + 6 + 6 = 24$

   $4 \times 6 = 24$

   To figure out the side lengths, I think "4 times what equals 24"? I know that $4 \times 6 = 24$, so each side is 6 centimeters.

   b. Find the area of Maureen's square.

   $6 \times 6 = 36$

   The area of Maureen's square is 36 square centimeters.

   I can multiply the side lengths to find the area.

   c. Estimate to draw a different rectangle that has the same perimeter as Maureen's square.

   Sample response:

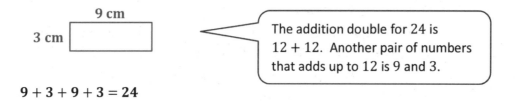

   9 cm

   3 cm

   The addition double for 24 is $12 + 12$. Another pair of numbers that adds up to 12 is 9 and 3.

   $9 + 3 + 9 + 3 = 24$

   d. Which shape has a greater area, Maureen's square or your rectangle?

   $3 \times 9 = 27$

   My rectangle has an area of 27 square centimeters. Maureen's square has a greater area because $36 > 27$.

   I can multiply 3×9 to find the area of my rectangle and then compare it to the area of Maureen's square.

Lesson 20:   Construct rectangles with a given perimeter using unit squares and
                             determine their areas.

EUREKA MATH

Name _____ Date _____

1. Cut out the unit squares at the bottom of the page. Then, use them to make as many rectangles as you can with a perimeter of 10 units.

    a. Estimate to draw your rectangles below. Label the side lengths of each rectangle.

    b. Find the areas of the rectangles in part (a) above.

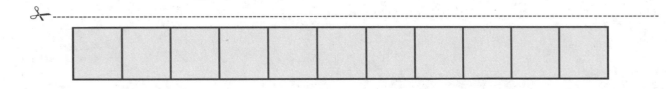

Lesson 20:   Construct rectangles with a given perimeter using unit squares and determine their areas.

© 2018 Great Minds®. eureka-math.org

251

2. Gino uses unit square tiles to make rectangles with a perimeter of 14 units. He draws his rectangles as shown below. Using square unit tiles, can Gino make another rectangle that has a perimeter of 14 units? Explain your answer.

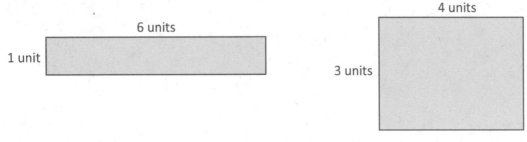

3. Katie draws a square that has a perimeter of 20 centimeters.

   a. Estimate to draw Katie's square below. Label the length and width of the square.

   b. Find the area of Katie's square.

   c. Estimate to draw a different rectangle that has the same perimeter as Katie's square.

   d. Which shape has a greater area, Katie's square or your rectangle?

Lesson 20: Construct rectangles with a given perimeter using unit squares and determine their areas.

© 2018 Great Minds®. eureka-math.org

253

1. Max uses unit squares to build rectangles that have a perimeter of 12 units. He creates the chart below to record his findings.

   a. Complete Max's chart. You might not use all the spaces in the chart.

   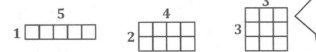

Perimeter = 12 units		
Number of rectangles I made: ___3___		
Width	**Length**	**Area**
1 unit	5 units	5 square units
2 units	*4 units*	*8 square units*
3 units	*3 units*	*9 square units*

   For a perimeter of ___ units, the total of all four side lengths has to be 12 units. I can think about the addition double for 12, which is $6 + 6$. That tells me that 6 units should be the sum of the length plus the width. I can find the same information by thinking about $12 \div 2$.

   To draw my rectangles, I think about pairs of numbers that equal 6 when I add them. The pairs I use to draw my rectangles are 1 and 5, 2 and 4, and 3 and 3. Then, to find the area of each rectangle, I multiply the side lengths. $1 \times 5 = 5$, $2 \times 4 = 8$, and $3 \times 3 = 9$. Now I can complete the chart.

   b. Explain how you found the widths and lengths in the chart above.

   I know that half of 12 is 6 because $6 + 6 = 12$. I thought about different ways to break apart 6. One way to break 6 apart is into 5 and 1. So, one rectangle can have side lengths of 5 units and 1 unit. Another way is 4 and 2. The last way to break apart 6 is 3 and 3. Those numbers became my side lengths.

Lesson 21:   Construct rectangles with a given perimeter using unit squares and
              determine their areas.

© 2018 Great Minds®. eureka-math.org

255

2. Grayson and Scarlett both draw rectangles with perimeters of 10 centimeters, but their rectangles have different areas. Explain with words, pictures, and numbers how this is possible.

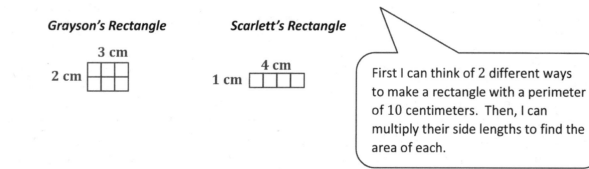

First I can think of 2 different ways to make a rectangle with a perimeter of 10 centimeters. Then, I can multiply their side lengths to find the area of each.

Grayson's and Scarlett's rectangles each have a perimeter of 10 centimeters. But the side lengths of their rectangles are different. That's what makes the product of the side lengths different, even though the sum is the same. The area of Grayson's rectangle is 6 square centimeters because $2 \times 3 = 6$. The area of Scarlett's rectangle is 4 square centimeters because $1 \times 4 = 4$.

Lesson 21:   Construct rectangles with a given perimeter using unit squares and
                         determine their areas.

Name _____   Date _____

1. Margo finds as many rectangles as she can with a perimeter of 14 centimeters.

    a. Shade Margo's rectangles on the grid below. Label the length and width of each rectangle.

    b. Find the areas of the rectangles in part (a) above.

    c. The perimeters of the rectangles are the same. What do you notice about the areas?

Lesson 21:   Construct rectangles with a given perimeter using unit squares and
determine their areas.

© 2018 Great Minds®. eureka-math.org

257

2. Tanner uses unit squares to build rectangles that have a perimeter of 18 units. He creates the chart below to record his findings.

   a. Complete Tanner's chart. You might not use all the spaces in the chart.

Perimeter = 18 units		
Number of rectangles I made: _____		
Width	**Length**	**Area**
1 unit	8 units	8 square units

   b. Explain how you found the widths and lengths in the chart above.

3. Jason and Dina both draw rectangles with perimeters of 12 centimeters, but their rectangles have different areas. Explain with words, pictures, and numbers how this is possible.

       Lesson 21:   Construct rectangles with a given perimeter using unit squares and determine their areas.

1. Jack uses square inch tiles to build a rectangle with a perimeter of 14 inches. Does knowing this help him find the number of rectangles he can build with an area of 14 square inches? Why or why not?

No, it doesn't. There is no connection between area and perimeter, so knowing how to build a rectangle with a perimeter of 14 inches doesn't help Jack figure out how many rectangles he can build with an area of 14 square inches.

I've studied area and perimeter a lot in class, and I know that they aren't related. If I want to know how many rectangles I can build with an area of 14 square inches, I can use square tiles or multiplication to figure it out. Thinking about perimeter won't help me.

2. Rachel makes a rectangle with a piece of string. She says the perimeter of her rectangle is 25 centimeters. Explain how it's possible for her rectangle's perimeter to be an odd number.

Most of the rectangles we've seen had an even perimeter because we usually look at rectangles with whole number side lengths. Rectangles can have odd perimeters if their side lengths are not whole numbers.

I know that rectangles with whole number side lengths have even perimeters because when you double the sum of whole numbers, you get an even number. Rectangles with fractional side lengths can have odd perimeters if the fractional parts add up to an odd number. For example, if a square has a side length of $\frac{1}{4}$, then the perimeter equals 1 because four copies of $\frac{1}{4}$ makes 1.

Lesson 22:   Use a line plot to record the number of rectangles constructed in Lessons 20 and 21.

259

Name _____ Date _____

1. The following line plot shows the number of rectangles a student made using square unit tiles. Use the line plot to answer the questions below.

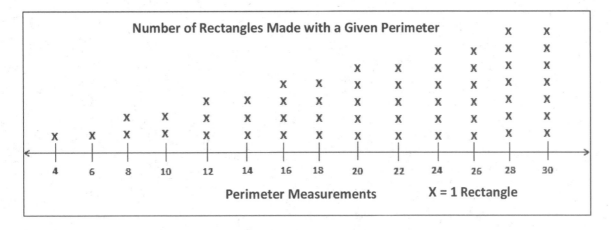

a. Why are all of the perimeter measurements even? Do all rectangles have even perimeters?

b. Explain the pattern in the line plot. What types of side lengths make this pattern possible?

c. How many X's would you draw for a perimeter of 32? Explain how you know.

Lesson 22:   Use a line plot to record the number of rectangles constructed in Lessons 20 and 21.

© 2018 Great Minds®. eureka-math.org

261

2. Luis uses square inch tiles to build a rectangle with a perimeter of 24 inches. Does knowing this help him find the number of rectangles he can build with an area of 24 square inches? Why or why not?

3. Esperanza makes a rectangle with a piece of string. She says the perimeter of her rectangle is 33 centimeters. Explain how it's possible for her rectangle to have an odd perimeter.

Lesson 22:   Use a line plot to record the number of rectangles constructed in
               Lessons 20 and 21.

© 2018 Great Minds®. eureka-math.org

1. Madison uses 4-inch square tiles to make a rectangle, as shown below. What is the perimeter of the rectangle in inches?

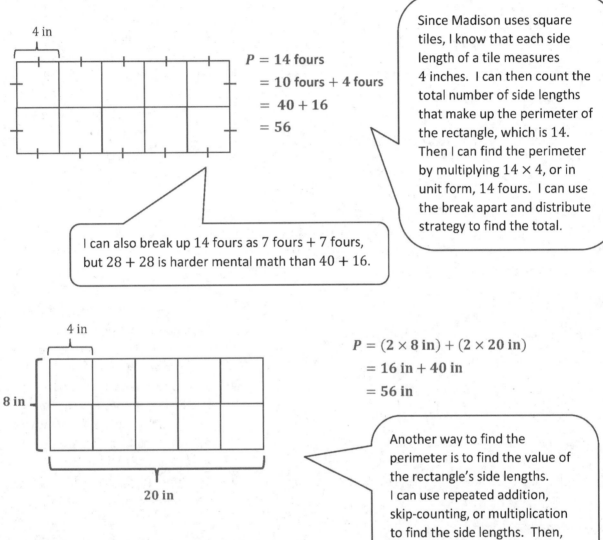

4 in

$P = $ **14 fours**

$= $ **10 fours + 4 fours**

$= $ **40 + 16**

$= $ **56**

Since Madison uses square tiles, I know that each side length of a tile measures 4 inches. I can then count the total number of side lengths that make up the perimeter of the rectangle, which is 14. Then I can find the perimeter by multiplying 14×4, or in unit form, 14 fours. I can use the break apart and distribute strategy to find the total.

I can also break up 14 fours as 7 fours + 7 fours, but $28 + 28$ is harder mental math than $40 + 16$.

4 in

8 in

20 in

$P = (2 \times 8 \text{ in}) + (2 \times 20 \text{ in})$

$= 16 \text{ in} + 40 \text{ in}$

$= 56 \text{ in}$

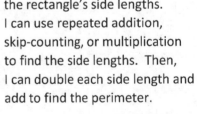

Another way to find the perimeter is to find the value of the rectangle's side lengths. I can use repeated addition, skip-counting, or multiplication to find the side lengths. Then, I can double each side length and add to find the perimeter.

The perimeter of the rectangle is 56 inches.

2. David traces 4 regular hexagons to create the shape shown below. The perimeter of 1 hexagon is 18 cm. What is the perimeter of David's new shape?

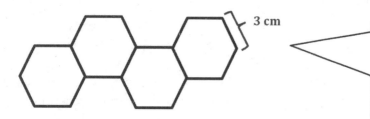

3 cm

> This is a two-step problem. First I need to find the side length of each hexagon. I know David traces regular hexagons, so all of the side lengths are equal. To find the side length, I can divide the perimeter of 1 hexagon, 18 cm, by its 6 sides to get 3 cm.

Perimeter of 1 hexagon = 18 cm ÷ 6

= 3 cm

Perimeter of the shape = 18 × 3 cm
$$= (10 \times 3 \text{ cm}) + (8 \times 3 \text{ cm})$$
$$= 30 \text{ cm} + 24 \text{ cm}$$
$$= 54 \text{ cm}$$

The perimeter of the shape is 54 cm.

> Next, I can count to find the total number of sides on David's new shape. I can't just multiply 4 × 6 to get the total number of sides because each hexagon shares 1 or 2 sides with another hexagon. I can mark the sides to help me count them. David's new shape has 18 sides. Now I can multiply 18 by 3 cm to get the perimeter of the shape.

Lesson 23:   Solve a variety of word problems with perimeter.

Name _____   Date _____

1. Rosie draws a square with a perimeter of 36 inches. What are the side lengths of the square?

2. Judith uses craft sticks to make two 24-inch by 12-inch rectangles. What is the total perimeter of the 2 rectangles?

3. An architect draws a square and a rectangle, as shown below, to represent a house that has a garage. What is the total perimeter of the house with its attached garage?

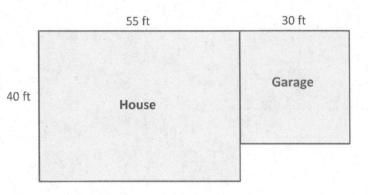

4. Manny draws 3 regular pentagons to create the shape shown below. The perimeter of 1 of the pentagons is 45 inches. What is the perimeter of Manny's new shape?

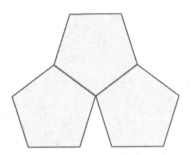

5. Johnny uses 2-inch square tiles to make a square, as shown below. What is the perimeter of Johnny's square?

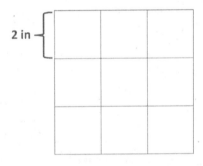

6. Lisa tapes three 7-inch by 9-inch pieces of construction paper together to make a happy birthday sign for her mom. She uses a piece of ribbon that is 144 inches long to make a border around the outside edges of the sign. How much ribbon is leftover?

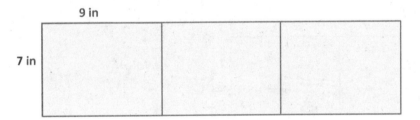

Lesson 23:   Solve a variety of word problems with perimeter.

1. Robin draws a square with a perimeter of 36 inches. What is the width and length of the square?

$36 \div 4 = 9$

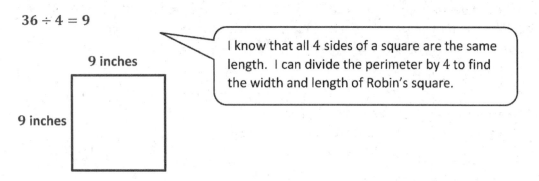

9 inches

9 inches

> I know that all 4 sides of a square are the same length. I can divide the perimeter by 4 to find the width and length of Robin's square.

The width and length of Robin's square are each 9 inches.

2. A rectangle has a perimeter of 16 centimeters.

   a. Estimate to draw as many different rectangles as you can that have a perimeter of 16 centimeters. Label the width and length of each rectangle.

   $16 \div 2 = 8$

   $1 + 7 = 8 \qquad w = 1, l = 7$
   $2 + 6 = 8 \qquad w = 2, l = 6$
   $3 + 5 = 8 \qquad w = 3, l = 5$
   $4 + 4 = 8 \qquad w = 4, l = 4$

   > I can divide the perimeter by 2 and then find pairs of numbers that have a sum of 8.

   7 cm

   1 cm

   6 cm

   2 cm

   5 cm

   3 cm

   4 cm

   4 cm

   > I can estimate to draw the 4 rectangles that I found.

Lesson 24:   Use rectangles to draw a robot with specified perimeter measurements, and reason about the different areas that may be produced.

© 2018 Great Minds®. eureka-math.org

267

b.   Explain the strategy you used to find the rectangles.

I divided the perimeter by 2, so $16 \div 2 = 8$. Then I found pairs of numbers that have a sum of 8. The pairs of numbers that have sums of 8 give me possible whole number side lengths for rectangles with a perimeter of 16 centimeters.

> I can divide the perimeter by 2 because the perimeter of a rectangle can be found by adding the width and the length and then multiplying by 2.
>
> $$\text{Perimeter} = 2 \times (\text{width} + \text{length})$$
> $$\text{Perimeter} \div 2 = \text{width} + \text{length}$$

Lesson 24:   Use rectangles to draw a robot with specified perimeter measurements, and reason about the different areas that may be produced.

© 2018 Great Minds®. eureka-math.org

Name _____ Date _____

1. Brian draws a square with a perimeter of 24 inches. What is the width and length of the square?

2. A rectangle has a perimeter of 18 centimeters.

   a. Estimate to draw as many different rectangles as you can that have a perimeter of 18 centimeters. Label the width and length of each rectangle.

   b. How many different rectangles did you find?

   c. Explain the strategy you used to find the rectangles.

Lesson 24:   Use rectangles to draw a robot with specified perimeter measurements, and reason about the different areas that may be produced.

© 2018 Great Minds®. eureka-math.org

269

3. The chart below shows the perimeters of three rectangles.

   a. Write possible widths and lengths for each given perimeter.

Rectangle	Perimeter	Width and Length
A	6 cm	_____ cm by _____ cm
B	10 cm	_____ cm by _____ cm
C	14 cm	_____ cm by _____ cm

   b. Double the perimeters of the rectangles in part (a). Then, find possible widths and lengths.

Rectangle	Perimeter	Width and Length
A	12 cm	_____ cm by _____ cm
B		_____ cm by _____ cm
C		_____ cm by _____ cm

Lesson 24:   Use rectangles to draw a robot with specified perimeter
              measurements, and reason about the different areas that may be
              produced.

EUREKA
MATH

The house below is made of rectangles and 1 triangle. The side lengths of each rectangle are labeled. Find the perimeter of each rectangle, and record it in the table on the next page.

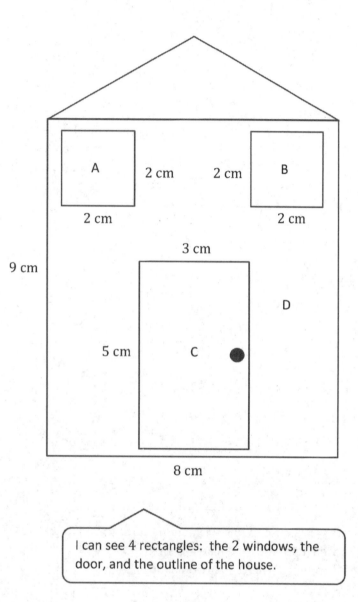

I can see 4 rectangles: the 2 windows, the door, and the outline of the house.

Rectangle	Perimeter
A	4×2 cm $= 8$ cm Perimeter $= 8$ cm
B	4×2 cm $= 8$ cm Perimeter $= 8$ cm
C	5 cm $+ 5$ cm $+ 3$ cm $+ 3$ cm $= 16$ cm Perimeter $= 16$ cm
D	8 cm $+ 8$ cm $+ 9$ cm $+ 9$ cm $= 34$ cm Perimeter $= 34$ cm

Rectangles A and B are squares, so I can find the perimeters by multiplying 4×2.

Another strategy I can use to find each perimeter is to add the width and length of the rectangle and then multiply the sum by 2. For Rectangle C, that would look like this:

$P = 2 \times (5 + 3)$

$P = 2 \times 8$

$P = 16$

Lesson 25: Use rectangles to draw a robot with specified perimeter measurements, and reason about the different areas that may be produced.

© 2018 Great Minds®. eureka-math.org

Name _____   Date _____

The robot below is made of rectangles. The side lengths of each rectangle are labeled. Find the perimeter of each rectangle, and record it in the table on the next page.

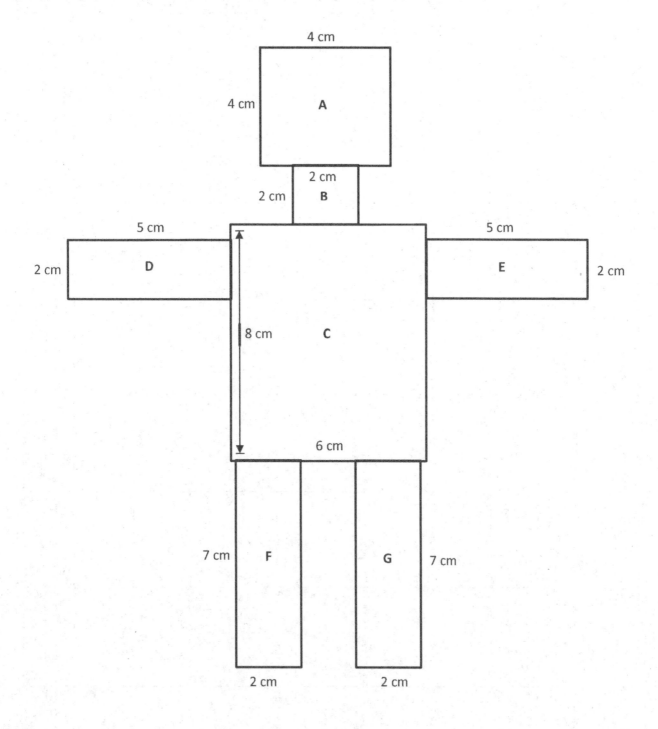

Rectangle	Perimeter
A	p = 4 × 4 cm P = 16 cm
B	
C	
D	
E	
F	
G	

Lesson 25: Use rectangles to draw a robot with specified perimeter measurements, and reason about the different areas that may be produced.

Each student in Mrs. William's class draws a rectangle with whole number side lengths and a perimeter of 32 centimeters. Then, they find the area of each rectangle and create the table below.

Area in Square Centimeters	Number of Students
15	1
28	2
39	2
48	3
55	4
60	6
63	2
64	2

> I know there can be many different areas for rectangles with the same perimeter.

a. What does this chart tell you about the relationship between area and perimeter?

The chart shows 8 different areas for rectangles with the same perimeter. So, I know that area and perimeter are 2 separate things. There's no connection between them.

b. Did any students in Mrs. William's class draw a square? Explain how you know.

Yes, 2 students drew a square. I know because I found all the possible side lengths of rectangles with a perimeter of 32 cm, and one rectangle has all equal side lengths of 8 cm. A square with side lengths of 8 cm has an area of 64 sq cm. On the chart, it shows that 2 students drew a rectangle with an area of 64 square centimeters.

> Perimeter is double the sum of the width and length of a rectangle. To find the side lengths of a rectangle with a perimeter of 32, I'll start by dividing the perimeter by 2 to get 16. Then, I can find pairs of numbers that add up to 16. Those are the possible side lengths.

c. What are the side lengths of the rectangle that most students in Mrs. William's class made?

I see that most students drew a rectangle with an area of 60 square centimeters. The side lengths of this rectangle are 6 cm and 10 cm.

Name _____   Date _____

1. Use Rectangles A and B to answer the questions below.

a. What is the perimeter of Rectangle A?

b. What is the perimeter of Rectangle B?

c. What is the area of Rectangle A?

d. What is the area of Rectangle B?

e. Use your answers to parts (a–d) to help you explain the relationship between area and perimeter.

2. Each student in Mrs. Dutra's class draws a rectangle with whole number side lengths and a perimeter of 28 centimeters. Then, they find the area of each rectangle and create the table below.

Area in Square Centimeters	Number of Students
13	2
24	1
33	3
40	5
45	4
48	2
49	2

a. Give two examples from Mrs. Dutra's class to show how it is possible to have different areas for rectangles that have the same perimeter.

b. Did any students in Mrs. Dutra's class draw a square? Explain how you know.

c. What are the side lengths of the rectangle that most students in Mrs. Dutra's class made with a perimeter of 28 centimeters?

Lesson 26: Use rectangles to draw a robot with specified perimeter measurements, and reason about the different areas that may be produced.

© 2018 Great Minds®. eureka-math.org

EUREKA MATH

Record the perimeters and areas of the rectangles in the chart on the next page.

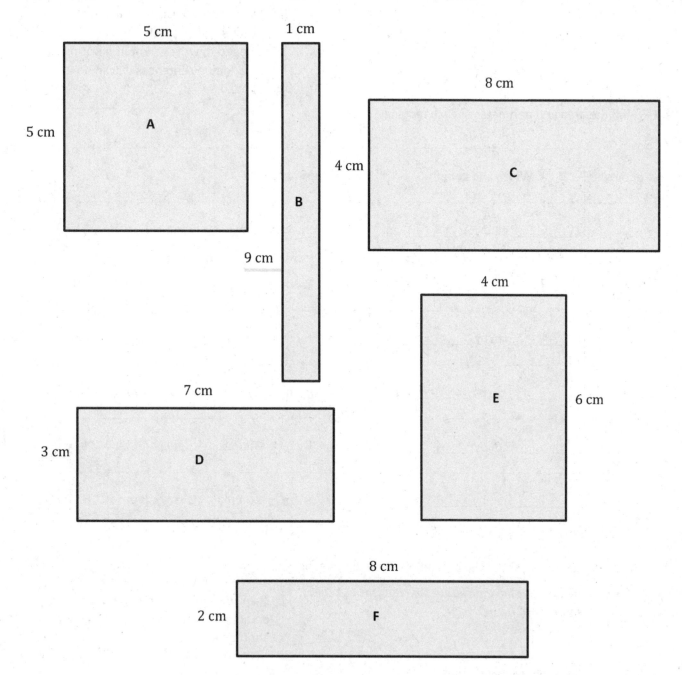

EUREKA MATH

Lesson 27: Use rectangles to draw a robot with specified perimeter measurements, and reason about the different areas that may be produced.

© 2018 Great Minds®. eureka-math.org

279

> I can choose to use mental math to solve for the perimeter and area. I do not need to write out multiplication and addition sentences if I can do it in my head.

1. Find the area and perimeter of each rectangle.

Rectangle	Width and Length	Perimeter	Area
A	____5____ cm by ____5____ cm	4×5 cm $= 20$ cm	5 cm $\times 5$ cm $= 25$ sq cm
B	____9____ cm by ____1____ cm	18 cm $+ 2$ cm $= 20$ cm	9 cm $\times 1$ cm $= 9$ sq cm
C	____4____ cm by ____8____ cm	8 cm $+ 16$ cm $= 20$ cm	4 cm $\times 8$ cm $= 32$ sq cm
D	____3____ cm by____7____ cm	6 cm $+ 14$ cm $= 20$ cm	3 cm $\times 7$ cm $= 21$ sq cm
E	____6____ cm by ____4____ cm	12 cm $+ 8$ cm $= 20$ cm	6 cm $\times 4$ cm $= 24$ sq cm
F	____2____ cm by____8____ cm	4 cm $+ 16$ cm $= 20$ cm	2 cm $\times 8$ cm $= 16$ sq cm

2. What do you notice about the perimeters of all the rectangles?

All of the rectangles have different side lengths but the same perimeter of 20 *cm.*

> I can see again how perimeter and area do not have any connection with one another.

3. Which rectangle is a square? How do you know?

Rectangle A is a square. I know because the width and length have the same measurement. Since opposite sides of rectangles are equal, Rectangle A has all equal side lengths and 4 right angles. That means it's a square!

Lesson 27:   Use rectangles to draw a robot with specified perimeter measurements, and reason about the different areas that may be produced.

EUREKA MATH

Name _____ Date _____

Record the perimeters and areas of the rectangles in the chart on the next page.

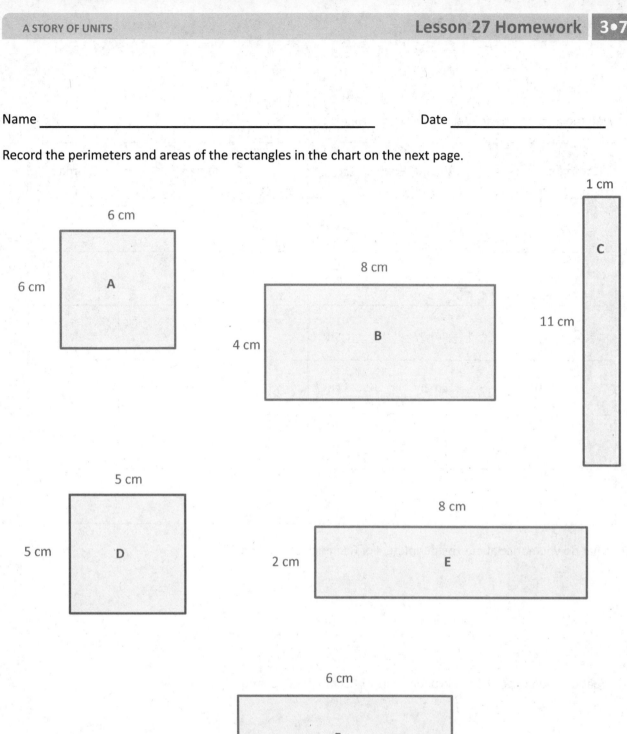

1 cm

6 cm

6 cm A

8 cm

4 cm B

C

11 cm

5 cm

5 cm D

8 cm

2 cm E

6 cm

4 cm F

1. Find the area and perimeter of each rectangle.

Rectangle	Width and Length	Perimeter	Area
A	_____ cm by _____ cm		
B	_____ cm by _____ cm		
C	_____ cm by _____ cm		
D	_____ cm by _____ cm		
E	_____ cm by _____ cm		
F	_____ cm by _____ cm		

2. What do you notice about the perimeters of Rectangles A, B, and C?

3. What do you notice about the perimeters of Rectangles D, E, and F?

4. Which two rectangles are squares? Which square has the greater perimeter?

Lesson 27:   Use rectangles to draw a robot with specified perimeter measurements, and reason about the different areas that may be produced.
© 2018 Great Minds®. eureka-math.org

A square sheet of construction paper has side lengths of 9 inches.

a. Estimate to draw the square sheet of paper, and label the side lengths.

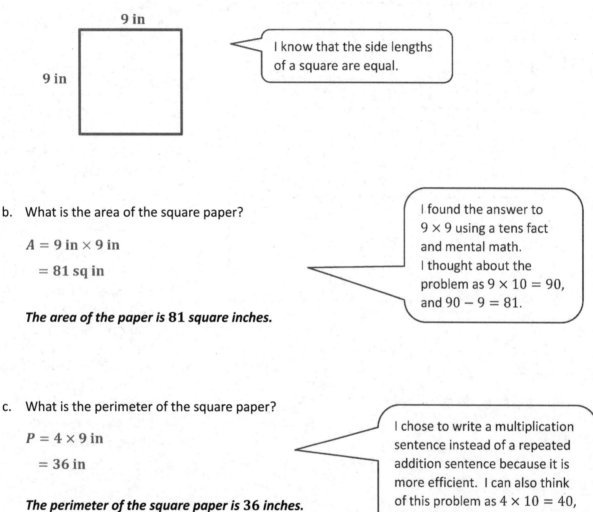

9 in

9 in

I know that the side lengths of a square are equal.

b. What is the area of the square paper?

$A = 9 \text{ in} \times 9 \text{ in}$

$= 81 \text{ sq in}$

The area of the paper is 81 square inches.

I found the answer to 9×9 using a tens fact and mental math. I thought about the problem as $9 \times 10 = 90$, and $90 - 9 = 81$.

c. What is the perimeter of the square paper?

$P = 4 \times 9 \text{ in}$

$= 36 \text{ in}$

The perimeter of the square paper is 36 inches.

I chose to write a multiplication sentence instead of a repeated addition sentence because it is more efficient. I can also think of this problem as $4 \times 10 = 40$, and $40 - 4 = 36$.

Lesson 28:   Solve a variety of word problems involving area and perimeter using all four operations.

© 2018 Great Minds®. eureka-math.org

d. Caitlyn connects three of these square papers to make one long banner. What is the perimeter of the new rectangular banner?

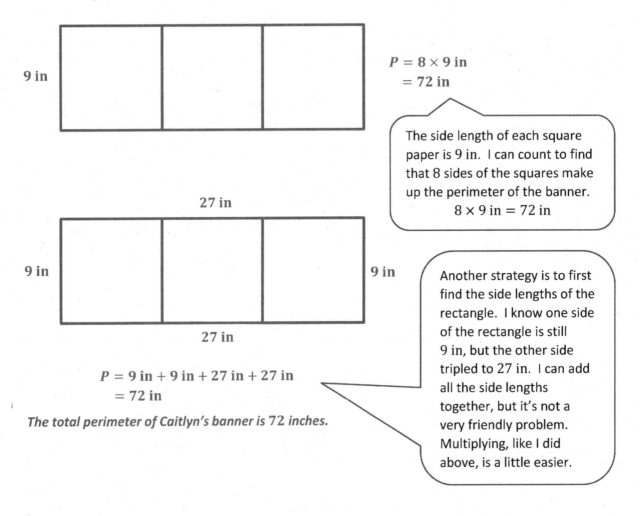

$P = 8 \times 9 \text{ in}$
$\quad = 72 \text{ in}$

> The side length of each square paper is 9 in. I can count to find that 8 sides of the squares make up the perimeter of the banner.
> $8 \times 9 \text{ in} = 72 \text{ in}$

$P = 9 \text{ in} + 9 \text{ in} + 27 \text{ in} + 27 \text{ in}$
$\quad = 72 \text{ in}$

The total perimeter of Caitlyn's banner is 72 inches.

> Another strategy is to first find the side lengths of the rectangle. I know one side of the rectangle is still 9 in, but the other side tripled to 27 in. I can add all the side lengths together, but it's not a very friendly problem. Multiplying, like I did above, is a little easier.

e. What is the total area of Caitlyn's banner?

$A = (3 \times 81 \text{ sq in})$
$\quad = (3 \times 80 \text{ sq in}) + (3 \times 1 \text{ sq in})$
$\quad = 240 \text{ sq in} + 3 \text{ sq in}$
$\quad = 243 \text{ sq in}$

> I can use the break apart and distribute strategy to help me find the answer to this challenging multiplication equation. I can first think of 3×80 in unit form as 3×8 tens $= 24$ tens, which has a value of 240. Then, I just have to remember to add the product of 3×1.

The total area of Caitlyn's banner is 243 square inches.

Lesson 28: Solve a variety of word problems involving area and perimeter using all four operations.

Name _____   Date _____

1. Carl draws a square that has side lengths of 7 centimeters.

   a. Estimate to draw Carl's square, and label the side lengths.

   b. What is the area of Carl's square?

   c. What is the perimeter of Carl's square?

   d. Carl draws two of these squares to make one long rectangle. What is the perimeter of this rectangle?

Lesson 28:   Solve a variety of word problems involving area and perimeter using all four operations.

© 2018 Great Minds®. eureka-math.org

285

2. Mr. Briggs puts food for the class party on a rectangular table. The table has a perimeter of 18 feet and a width of 3 feet.

    a. Estimate to draw the table, and label the side lengths.

    b. What is the length of the table?

    c. What is the area of the table?

    d. Mr. Briggs puts three of these tables together side by side to make 1 long table. What is the area of the long table?

Lesson 28:   Solve a variety of word problems involving area and perimeter using all four operations.

Josh puts two rectangles together to make the L-shaped figure below. He measures some of the side lengths and records them as shown.

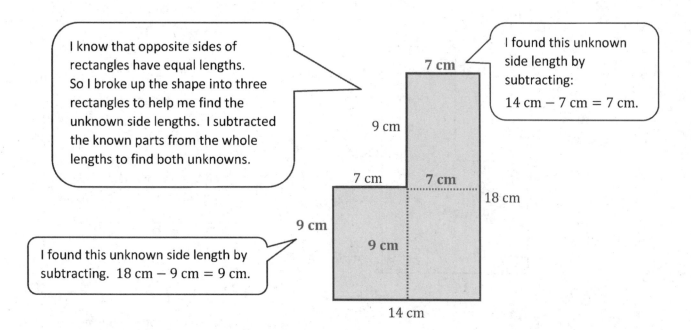

I know that opposite sides of rectangles have equal lengths. So I broke up the shape into three rectangles to help me find the unknown side lengths. I subtracted the known parts from the whole lengths to find both unknowns.

I found this unknown side length by subtracting:
14 cm − 7 cm = 7 cm.

I found this unknown side length by subtracting. 18 cm − 9 cm = 9 cm.

a.   Find the perimeter of Josh's shape.

$$P = (2 \times 18 \text{ cm}) + (2 \times 14 \text{ cm})$$
$$= 36 \text{ cm} + 28 \text{ cm}$$
$$= 64 \text{ cm}$$

The perimeter of Josh's shape is 64 cm.

Lesson 29:   Solve a variety of word problems involving area and perimeter using all four operations.

287

© 2018 Great Minds®. eureka-math.org

b. Find the area of Josh's shape.

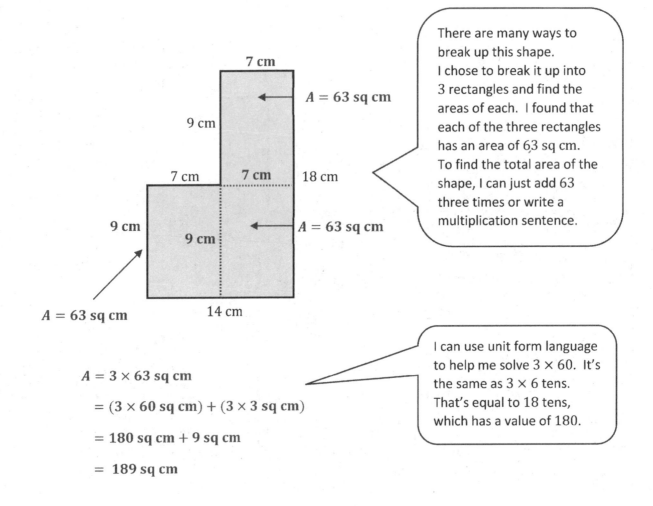

There are many ways to break up this shape. I chose to break it up into 3 rectangles and find the areas of each. I found that each of the three rectangles has an area of 63 sq cm. To find the total area of the shape, I can just add 63 three times or write a multiplication sentence.

I can use unit form language to help me solve 3×60. It's the same as 3×6 tens. That's equal to 18 tens, which has a value of 180.

$$A = 3 \times 63 \text{ sq cm}$$

$$= (3 \times 60 \text{ sq cm}) + (3 \times 3 \text{ sq cm})$$

$$= 180 \text{ sq cm} + 9 \text{ sq cm}$$

$$= 189 \text{ sq cm}$$

The area of Josh's shape is **189 sq cm.**

Lesson 29:   Solve a variety of word problems involving area and perimeter using all four operations.

© 2018 Great Minds®. eureka-math.org

EUREKA MATH®

Name _____ Date _____

1. Katherine puts two squares together to make the rectangle below. The side lengths of the squares measure 8 inches.

8 in

a. What is the perimeter of the rectangle Katherine made with her 2 squares?

b. What is the area of Katherine's rectangle?

c. Katherine decides to draw another rectangle of the same size. What is the area of the new, larger rectangle?

8 in

Lesson 29:   Solve a variety of word problems involving area and perimeter using all four operations.

289

© 2018 Great Minds®. eureka-math.org

2. Daryl draws 6 equal-sized rectangles as shown below to make a new, larger rectangle. The area of one of the small rectangles is 12 square centimeters, and the width of the small rectangle is 4 centimeters.

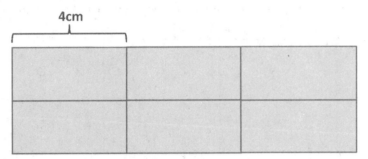

a. What is the perimeter of Daryl's new rectangle?

b. What is the area of Daryl's new rectangle?

3. The recreation center soccer field measures 35 yards by 65 yards. Chris dribbles the soccer ball around the perimeter of the field 4 times. What is the total number of yards Chris dribbles the ball?

Lesson 29:   Solve a variety of word problems involving area and perimeter using all four operations.

Andrew solves the following problem as shown below.

A basketball court measures 74 feet by 52 feet. Bill dribbles the basketball around the court sidelines 3 times. What is the total number of feet Bill dribbles the ball?

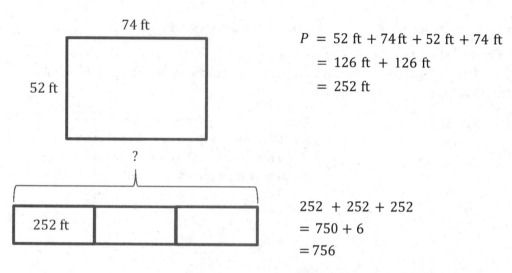

$P = 52 \text{ ft} + 74 \text{ ft} + 52 \text{ ft} + 74 \text{ ft}$

$= 126 \text{ ft} + 126 \text{ ft}$

$= 252 \text{ ft}$

$252 + 252 + 252$

$= 750 + 6$

$= 756$

Bill dribbles the ball 756 feet.

1. What strategies did Andrew use to solve this problem?

    Andrew drew a picture of the basketball court and labeled the side lengths. Then he added to find the perimeter. Finally, he used a tape diagram to find the total of 3 perimeters.

> Analyzing my classmates' work improves my problem-solving skills because I am able to see different and sometimes more efficient ways of solving a problem.

2. What did Andrew do well?

    Andrew used all the steps in the RDW process. He used mental math for his calculations. He also drew and labeled a tape diagram to show his thinking for his second step.

3. What are some suggestions that you would give Andrew to improve his work?

   Some suggestions would be to have Andrew use a letter to represent the unknown in the tape diagram and label all of the units in his addition sentence.

4. What are some strategies you would like to try based on Andrew's work?

   I would like to practice thinking about numbers like $252 + 252 + 252$ as $(250 + 250 + 250) + (2 + 2 + 2)$. That will help me use mental math strategies to add and not have to use the algorithm as much.

   > Having classmates analyze my work is helpful because I am able to get ideas on how to improve it.

   Lesson 30:   Share and critique peer strategies for problem solving.

Name _____ Date _____

Use this form to critique Student A's problem-solving work on the next page.

Student:	Student A	Problem Number:	
Strategies Student A Used:			
Things Student A Did Well:			
Suggestions for Improvement:			
Strategies I Would Like to Try Based on Student A's Work:			

Lesson 30:   Share and critique peer strategies for problem solving.

© 2018 Great Minds®. eureka-math.org

293

Name _____ **STUDENT A** _____   Date _____

1. Katherine puts 2 squares together to make the rectangle below. The side lengths of the squares measure 8 inches.

8 in

a. What is the perimeter of Katherine's rectangle?

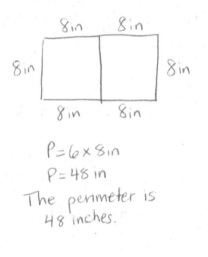

$P = 6 \times 8$ in
$P = 48$ in
The perimeter is 48 inches.

b. What is the area of Katherine's rectangle?

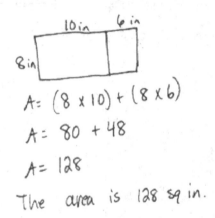

$A = (8 \times 10) + (8 \times 6)$
$A = 80 + 48$
$A = 128$
The area is 128 sq in.

EUREKA
MATH

c. Katherine draws 2 of the rectangles in Problem 1 side by side. Her new, larger rectangle is shown below. What is the area of the new, larger rectangle?

8 in

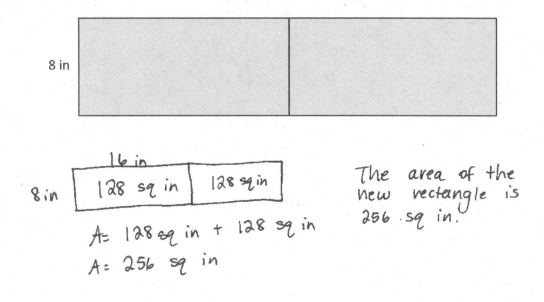

16 in

8 in | 128 sq in | 128 sq in

A = 128 sq in + 128 sq in
A = 256 sq in

The area of the new rectangle is 256 sq in.

1. Use the rectangle below to answer Problem 1 (a)–(d).

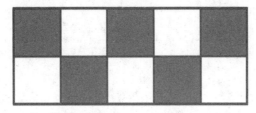

a. What is the area of the rectangle in square units?

    The area of the rectangle is 10 square units.

    > I can find the area by multiplying the side lengths.
    > $2 \times 5 = 10$
    > Or, I can count the square units. Either way the answer is the same!

b. What is the area of half of the rectangle in square units?

    $10 \div 2 = 5$

    > I can divide the total area by 2 to find the area of half of the rectangle.

    The area of half of the rectangle is 5 square units.

c. Shade in half of the rectangle above. Be creative with your shading!

    > I can use my answer to part (b) to help me shade in half of the rectangle.

d. Explain how you know you shaded in half of the rectangle.

    I know I shaded in half of the rectangle because I shaded 5 square units and the area of half of the rectangle is 5 square units.

2. During art class, Mia draws a shape and then shades one-half of it. Analyze Mia's work. Determine if she was correct or not, and explain your thinking.

Mia's Drawing	Your Analysis
	Mia did not correctly shaded one-half of her drawing. There is less than one-half of the drawing shaded because of the unshaded heart in the shaded part of the drawing. She needs to shade a same-sized heart in the unshaded part to show one-half shaded.

I can picture what Mia's drawing might look like if she had shaded it correctly. It might look like this:

3. Shade the grid below to show two different ways of shading half of each shape.

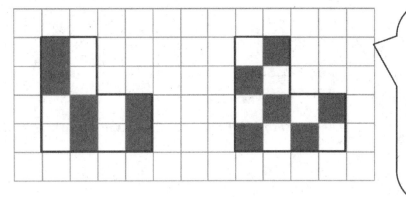

I can find the total area for each shape by counting the square units. Then I can divide that number by 2 to figure out how many square units to shade in order to show one-half. I can shade in 6 square units for each shape.

$12 \div 2 = 6$

Lesson 31: Explore and create unconventional representations of one-half.

EUREKA MATH®

Name _____   Date _____

1. Use the rectangle below to answer Problem 1(a–d).

a. What is the area of the rectangle in square units?

b. What is the area of half of the rectangle in square units?

c. Shade in half of the rectangle above. Be creative with your shading!

d. Explain how you know you shaded in half of the rectangle.

2. During math class, Arthur, Emily, and Gia draw a shape and then shade one-half of it. Analyze each student's work. Determine if each student was correct or not, and explain your thinking.

Student	Drawing	Your Analysis
Arthur		
Emily		
Gia		

3. Shade the grid below to show two different ways of shading half of each shape.

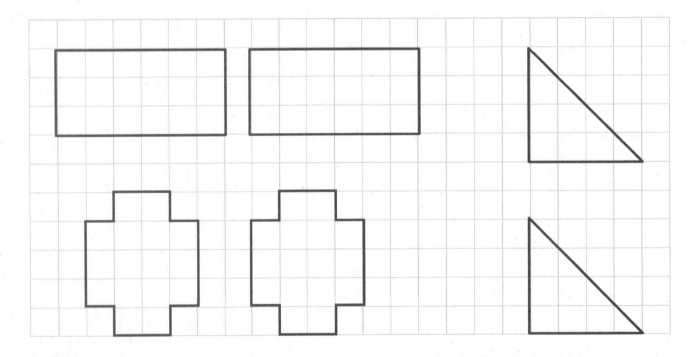

    Lesson 31:   Explore and create unconventional representations of one-half.

1. Estimate to finish shading the circle below so that it is about one-half shaded.

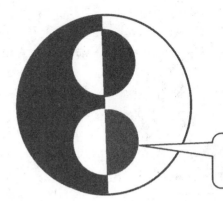

I can shade in another half circle that is about the same size as the unshaded half circle.

2. Explain how you know the circle in Problem 1 is about one-half shaded.

I know the circle in Problem 1 is about one-half shaded because I can picture the little shaded half circles flipped over and moved into the shaded part of the circle. Then it would be easy to see that the circle is about one-half shaded because it would look like this:

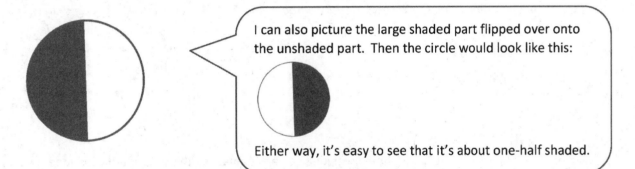

I can also picture the large shaded part flipped over onto the unshaded part. Then the circle would look like this:

Either way, it's easy to see that it's about one-half shaded.

3. Can you say the circle in Problem 1 is exactly one-half shaded? Why or why not?

No, I can't say that the circle in Problem 1 is exactly one-half shaded because there aren't any gridlines, and I had to estimate to shade the little half circle. When I estimate, I know my answer isn't exact.

I can also tell the circle is not exactly one-half shaded because the directions for Problems 1 and 2 use the word *about*. When I see the word *about* I know the answer is not exact; it's an estimate.

Lesson 32:   Explore and create unconventional representations of one-half.

301

© 2018 Great Minds®. eureka-math.org

4. Wilson and Laurie shade in circles as shown below.

Wilson's Circle   **Laurie's Circle**

a. Whose circle is about one-half shaded? How do you know?

Laurie's circle is about one-half shaded. I can picture the image in the top part of the circle flipped over and moved to the bottom of the circle. Then the bottom half of Laurie's circle would be all shaded, which means the whole circle would be about one-half shaded.

> I see that the shaded amount is about the same as the unshaded amount in Laurie's circle. That means that Laurie's circle is about one-half shaded.

b. Explain how the circle that is not one-half shaded can be changed so that it is one-half shaded.

Wilson's circle has too much shading. He needs to erase a small circle in one of the shaded parts that matches the small shaded circle.

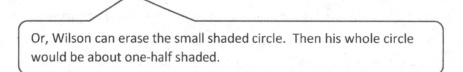

> Or, Wilson can erase the small shaded circle. Then his whole circle would be about one-half shaded.

Lesson 32:   Explore and create unconventional representations of one-half.

© 2018 Great Minds®. eureka-math.org

Name _____ Date _____

1. Estimate to finish shading the circles below so that each circle is about one-half shaded.

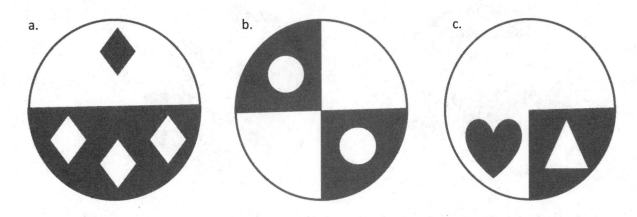

a.   b.   c.

2. Choose one of the circles in Problem 1, and explain how you know it's about one-half shaded.

    Circle _____

3. Can you say the circles in Problem 1 are exactly one-half shaded? Why or why not?

Lesson 32:   Explore and create unconventional representations of one-half.

303

© 2018 Great Minds®. eureka-math.org

4. Marissa and Jake shade in circles as shown below.

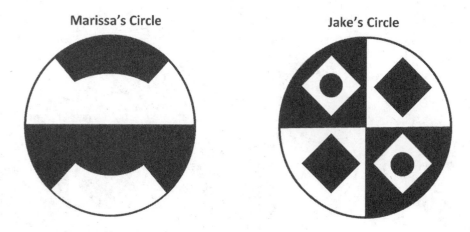

Marissa's Circle   Jake's Circle

a. Whose circle is about one-half shaded? How do you know?

b. Explain how the circle that is not one-half shaded can be changed so that it is one-half shaded.

5. Estimate to shade about one-half of each circle below in an unusual way.

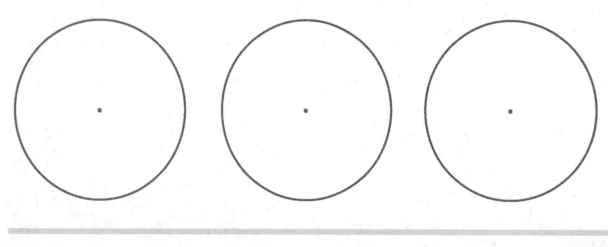

Lesson 32:   Explore and create unconventional representations of one-half.

© 2018 Great Minds®. eureka-math.org

Teach a family member your favorite fluency game from class. Record information about the game you taught below.

Name of the game:

Partition Shapes

Materials used:

The only materials we needed were personal white boards and markers.

> I can pick any activity from the list my teacher gave me and teach it to someone at home. I know how to play the game by myself, but sometimes you learn something by teaching it to someone else. It helped me think about fractions more when I had to show my sister what we needed to do.

Name of the person you taught to play:

I taught my sister Sonia to play.

Describe what it was like to teach the game. Was it easy? Hard? Why?

I'm used to learning games from my teacher and then playing with friends. Teaching someone else was fun, but it was tricky. Even though I know how to play the game, I realized after we started that I forgot to explain some of the important parts.

Will you play the game together again? Why or why not?

Yes. We liked drawing shapes on our personal white boards. My sister didn't know about fractions, and I got to show her. I liked that. We'll try different games sometimes too.

Was the game as fun to play at home as in class? Why or why not?

It was really fun to play at home because I also got to teach it to my sister.

Lesson 33: Solidify fluency with Grade 3 skills.

305

Name _____   Date _____

Teach a family member your favorite fluency game from class. Record information about the game you taught below.

Name of the game: _____

_____

Materials used: _____

_____

Name of the person you taught to play: _____

Describe what it was like to teach the game. Was it easy? Hard? Why? _____

_____

_____

_____

Will you play the game together again? Why or why not? _____

_____

_____

Was the game as fun to play at home as in class? Why or why not? _____

_____

_____

Credits

Great Minds® has made every effort to obtain permission for the reprinting of all copyrighted material. If any owner of copyrighted material is not acknowledged herein, please contact Great Minds for proper acknowledgment in all future editions and reprints of this module.